Big Ideas in Mathematics
for Future Mathematics Teachers

Big Ideas in Probability and Statistics

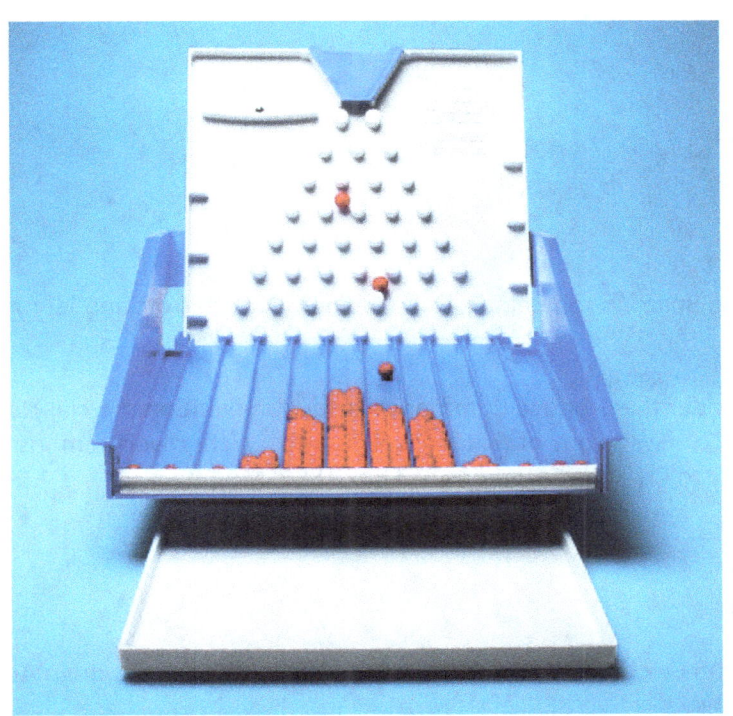

John Beam, Jason Belnap, Eric Kuennen,
Amy Parrott, and Jennifer Szydlik
(Updated Summer 2023)

Copyright 2021 by John Beam, Jason Belnap, Eric Kuennen, Amy Parrott and Jennifer Szydlik

This work is licensed under the Creative Commons Attribution-NonCommercial-NoDerivatives 4.0 International License. To view a copy of this license, visit http://creativecommons.org/licenses/by-nc-nd/4.0/ or send a letter to Creative Commons, PO Box 1866, Mountain View, CA 94042, USA.

Cover of photo of Galton Box made by Estes Objethos Atelier, photo by Rodrigo Argenton, CC-BY-SA 4.0 via Wikimedia Commons

Title page photo of Binostat by Invicta Education.

First Edition 2021

ISBN: 9798526758611

Eric W. Kuennen, Mathematics Department, University of Wisconsin Oshkosh
800 Algoma Blvd, Oshkosh, WI 54901, Mountain View, CA 94042, USA.

Dear Future Teacher,

We wrote this book to help you to see the structure that underlies elementary mathematics, to give you experiences really *doing* mathematics, and to show you how children think and learn. We fully intend this course to transform your relationship with math.

As teachers of future elementary teachers, we created or gathered the activities for this text, and then we tried them out with our own students and modified them based on their suggestions and insights. We know that some of the problems are tough – you will get stuck sometimes. Please don't let that discourage you. There's much value in wrestling with an idea.

                                            All our best,

                                            John, Jason, Eric, Amy & Jen

Table of Contents

Chance favors only the prepared mind.
Louis Pasteur

Hey! Read this. It will help you understand the book. 1

    Common Core State Standards 3

Chapter One: Conceptual Foundations

Class Activity 1a: Feet and Hands 10

Class Activity 1b: Tuition and Minimum Wage 11

    Read and Study: The Language of Statistics 12
        Scatterplots   13
        Association and Correlation 15

Class Activity 2: HIV and AZT 19

    Read and Study: Relationships between Categorical Variables 20
        The Philosophy of Statistics 20
        Sources of Bias   20

Class Activity 3: It's a Long Shot 26

    Read and Study: Concepts of Probability 27
        Probability Models 27
        Theoretical and Experimental Probability 28
    Connections to Teaching: Quantifying Chance 28

Class Activity 4: Spaghetti Triangles 33

    Read and Study: Randomness 35
       Data Displays   37
    Connections to Teaching: Probability Fallacies 38

Class Activity 5a: The Best Answer 43

Class Activity 5b: Hanging in the Balance 44

    Read and Study: Measures of Center 45
        Mean, Median, and Mode 45
        Sample Mean and Population Mean 46
        Connections to Teaching: Relating Mean and Median 47

Class Activity 6a: The St. Petersburg Lottery **50**

Class Activity 6b: The Problem of Points **51**

    Read and Study: Expected Value 52
        Fair Games   52
        Analyzing Games of Chance 53

Class Activity 7a: Shufflehall **55**

Class Activity 7b: How Many Tanks? **56**

    Read and Study: Measures of Spread 57
        Range, Interquartile Range, and Box Plots 57
        Standard Deviation 58
        Mean Absolute Deviation 59
    Connections to Teaching: Determining Outliers 59

Class Activity 8a: The Illuminated Dartboard **66**

Class Activity 8b: Three Prisoners **68**

    Read and Study: Conditional Probability 69
        Dependence and Independence 69
        Compound Events and Tree Diagrams 71

Chapter Two: Counting

Class Activity 9: Seating Arrangements **77**

    Read and Study: Permutations 78
        Counting Trees   78

Class Activity 10: M I S S I S S I P P I **83**

    Read and Study: Multiset Permutations 84

Class Activity 11: Pascal's Triangle **87**

    Read and Study: Combinations 88
        Patterns in Pascal's Triangle 89
        Relating Combinations, Permuatations and Multiset Permutations 90

Class Activity 12: Gimme Five! **92**

    Read and Study: The Binomial Distribution 93
        Calculating Binomial Probabilities 95

Class Activity 13a: All That and a Bag of Chips 97

Class Activity 13b: A Dicey Situation 98

    Read and Study: Sampling Distributions..99
        The Law of Large Numbers..100
        The Central Limit Theorem 102

Chapter Three ;Inference

Class Activity 14: Which Bag is Which? 106

    Read and Study: Introduction to Inference 107

Class Activity 15a: A Loaded Question..110

Class Activity 15b: ESP Testing 111

    Read and Study: Hypothesis Testing...112
        Levels of Significance 112
        Comparing Two Proportions 113
    Read and Study: The Normal Distribution..119

Class Activity 16: Yanny or Laurel? 122

    Read and Study: Confidence Intervals 123
        Estimating the Population Proportion..123

Class Activity 17: Fingerprints 127

    Read and Study: Bayes' Theorem 128
        Frequency Analysis...128
        More Conditional Probability Problems 129

References 134

Glossary 140

Hey! Read this. It will help you understand the book.

The only way to learn mathematics is to do mathematics.
                                                    Paul Halmos

This book was written to prepare future math teacher for the mathematical work of teaching. The focus of this module is probability and statistics – areas that have received increased attention in the middle grades during the past decade, and for good reason. Statistics is important for understanding politics, science, and economics. In fact, the National Council of Teachers of Mathematics (NCTM) asserts, "Students need to know about data analysis and related aspects of probability in order to reason statistically – skills necessary to becoming informed citizens and intelligent consumers" (p. 48), and recommends that an increased emphasis on these ideas span all grade levels. (NCTM, 2000). The ideas in this book are fundamentally important for your students to understand and so they are fundamentally important for *you* to understand.

As mathematicians, we will also try to convey to you the beauty of our subject. Mathematicians view mathematics as the study of patterns and structures. We want to show you how to reason like a mathematician – and we want you to show this to your students too. This *way of reasoning* is just as important as any content you teach. When you stand before your class, you are a representative of the mathematical community; we will help you to become a good one. No one can do this thinking for you. Mathematics isn't a subject you can memorize; it is about ways of thinking and knowing. *You* need to do examples, gather data, look for patterns, experiment, draw pictures, think, try again, make arguments, and think some more. The big ideas of probability and statistics are not always easy.

Each section of this book begins with a **Class Activity**. The activity is designed for small-group work in class. Some activities may take your class as little as 30 minutes to complete and discuss. Others may take you two or more class periods. No solutions are provided to activities – you will have to solve them yourselves. The **Read and Study**, **Connections to Teaching**, and **Homework** sections are presented within the context of the activity ideas.

In preparing to write this text, we studied four *Standards*-based curriculum projects for middle school students (the books your future students might use). Those projects are *Mathematics in Context*, *Connected Mathematics*, *MATHThematics*, and *MathScape*. All of these are activity-based and *Standards*-based curricula. This means that the middle school materials were written so that your future students will solve problems and create understandings based on concrete experiences. In case you are skeptical about these types of materials for your future students, let us assure you that they better encourage and support the types of behaviors and thinking that mathematicians value than do traditional materials. Furthermore, the research suggests that schools that adopt *Standards*-based materials for more than two years show significantly higher test scores on even traditional measures of mathematical understanding than matched schools that adopt traditional curricula (Reys, Reys, Lapan, & Holliday, 2003; Riordan & Noyce, 2001; Briars, 2001; Griffen, Evans, Timms, & Trowell, 2000; Mullis et al., 2001). We assure you that the ideas you will meet in these pages are vitally connected to the mathematics curriculum of your future students, and we hope that the text is written in a way that makes these connections apparent to you.

The Common Core State Standards is "a state-led effort to establish a shared set of clear educational standards for English language arts and mathematics that states can voluntarily adopt. The standards have been informed by the best available evidence and the highest state standards across the country and globe and designed by a diverse group of teachers, experts, parents, and school administrators…" (see http://www.corestandards.org/Math/) As of the time of publication of this text, most states had officially adopted these standards, and so it is important for you to know them and the content and practices that they advocate.

On the following pages, we have reproduced the Common Core State Standards for Probability and Statistics in grades 6-8. Moreover, the Common Core Standards for Mathematical Practice have articulated what it means to "do" mathematics, regardless of the content domain. Throughout the text we will ask you to refer back to these standards to reflect upon.

Common Core State Standards

Statistics and Probability

Grade 6

Develop understanding of statistical variability.

1. Recognize a statistical question as one that anticipates variability in the data related to the question and accounts for it in the answers. *For example, "How old am I?" is not a statistical question, but "How old are the students in my school?" is a statistical question because one anticipates variability in students' ages.*

2. Understand that a set of data collected to answer a statistical question has a distribution which can be described by its center, spread, and overall shape.

3. Recognize that a measure of center for a numerical data set summarizes all of its values with a single number, while a measure of variation describes how its values vary with a single number.

Summarize and describe distributions.

4. Display numerical data in plots on a number line, including dot plots, histograms, and box plots.

5. Summarize numerical data sets in relation to their context, such as by:

    a. Reporting the number of observations.

    b. Describing the nature of the attribute under investigation, including how it was measured and its units of measurement.

    c. Giving quantitative measures of center (median and/or mean) and variability (interquartile range and/or mean absolute deviation), as well as describing any overall pattern and any striking deviations from the overall pattern with reference to the context in which the data were gathered.

    d. Relating the choice of measures of center and variability to the shape of the data distribution and the context in which the data were gathered.

Grade 7

Use random sampling to draw inferences about a population.

1. Understand that statistics can be used to gain information about a population by examining a sample of the population; generalizations about a population from a sample are valid only if the sample is representative of that population. Understand that random sampling tends to produce representative samples and support valid inferences.

2. Use data from a random sample to draw inferences about a population with an unknown characteristic of interest. Generate multiple samples (or simulated samples) of the same size to gauge the variation in estimates or predictions. *For example, estimate the mean word length in a book by randomly sampling words from the book; predict the winner of a school election based on randomly sampled survey data. Gauge how far off the estimate or prediction might be.*

© Copyright 2010. National Governors Association Center for Best Practices and Council of Chief State School Officers. All rights reserved.

Draw informal comparative inferences about two populations.

3 Informally assess the degree of visual overlap of two numerical data distributions with similar variabilities, measuring the difference between the centers by expressing it as a multiple of a measure of variability. *For example, the mean height of players on the basketball team is 10 cm greater than the mean height of players on the soccer team, about twice the variability (mean absolute deviation) on either team; on a dot plot, the separation between the two distributions of heights is noticeable.*

4 Use measures of center and measures of variability for numerical data from random samples to draw informal comparative inferences about two populations. *For example, decide whether the words in a chapter of a seventh-grade science book are generally longer than the words in a chapter of a fourth-grade science book.*

Investigate chance processes and develop, use, and evaluate probability models.

5 Understand that the probability of a chance event is a number between 0 and 1 that expresses the likelihood of the event occurring. Larger numbers indicate greater likelihood. A probability near 0 indicates an unlikely event, a probability around 1/2 indicates an event that is neither unlikely nor likely, and a probability near 1 indicates a likely event.

6 Approximate the probability of a chance event by collecting data on the chance process that produces it and observing its long-run relative frequency, and predict the approximate relative frequency given the probability. *For example, when rolling a number cube 600 times, predict that a 3 or 6 would be rolled roughly 200 times, but probably not exactly 200 times.*

7 Develop a probability model and use it to find probabilities of events. Compare probabilities from a model to observed frequencies; if the agreement is not good, explain possible sources of the discrepancy.

   a Develop a uniform probability model by assigning equal probability to all outcomes, and use the model to determine probabilities of events. *For example, if a student is selected at random from a class, find the probability that Jane will be selected and the probability that a girl will be selected.*

   b Develop a probability model (which may not be uniform) by observing frequencies in data generated from a chance process. *For example, find the approximate probability that a spinning penny will land heads up or that a tossed paper cup will land open-end down. Do the outcomes for the spinning penny appear to be equally likely based on the observed frequencies?*

© Copyright 2010. National Governors Association Center for Best Practices and Council of Chief State School Officers. All rights reserved.

8 Find probabilities of compound events using organized lists, tables, tree diagrams, and simulation.

   a Understand that, just as with simple events, the probability of a compound event is the fraction of outcomes in the sample space for which the compound event occurs.

   b Represent sample spaces for compound events using methods such as organized lists, tables and tree diagrams. For an event described in everyday language (e.g., "rolling double sixes"), identify the outcomes in the sample space which compose the event.

   c Design and use a simulation to generate frequencies for compound events. *For example, use random digits as a simulation tool to approximate the answer to the question: If 40% of donors have type A blood, what is the probability that it will take at least 4 donors to find one with type A blood?*

Grade 8
Investigate patterns of association in bivariate data.

1 Construct and interpret scatter plots for bivariate measurement data to investigate patterns of association between two quantities. Describe patterns such as clustering, outliers, positive or negative association, linear association, and nonlinear association.

2 Know that straight lines are widely used to model relationships between two quantitative variables. For scatter plots that suggest a linear association, informally fit a straight line, and informally assess the model fit by judging the closeness of the data points to the line.

3 Use the equation of a linear model to solve problems in the context of bivariate measurement data, interpreting the slope and intercept. *For example, in a linear model for a biology experiment, interpret a slope of 1.5 cm/hr as meaning that an additional hour of sunlight each day is associated with an additional 1.5 cm in mature plant height.*

4 Understand that patterns of association can also be seen in bivariate categorical data by displaying frequencies and relative frequencies in a two-way table. Construct and interpret a two-way table summarizing data on two categorical variables collected from the same subjects. Use relative frequencies calculated for rows or columns to describe possible association between the two variables. *For example, collect data from students in your class on whether or not they have a curfew on school nights and whether or not they have assigned chores at home. Is there evidence that those who have a curfew also tend to have chores?*

© Copyright 2010. National Governors Association Center for Best Practices and Council of Chief State School Officers. All rights reserved.

Mathematics | Standards for Mathematical Practice

The Standards for Mathematical Practice describe varieties of expertise that mathematics educators at all levels should seek to develop in their students. These practices rest on important "processes and proficiencies" with longstanding importance in mathematics education. The first of these are the NCTM process standards of problem solving, reasoning and proof, communication, representation, and connections. The second are the strands of mathematical proficiency specified in the National Research Council's report *Adding It Up*: adaptive reasoning, strategic competence, conceptual understanding (comprehension of mathematical concepts, operations and relations), procedural fluency (skill in carrying out procedures flexibly, accurately, efficiently and appropriately), and productive disposition (habitual inclination to see mathematics as sensible, useful, and worthwhile, coupled with a belief in diligence and one's own efficacy).

1 Make sense of problems and persevere in solving them.

Mathematically proficient students start by explaining to themselves the meaning of a problem and looking for entry points to its solution. They analyze givens, constraints, relationships, and goals. They make conjectures about the form and meaning of the solution and plan a solution pathway rather than simply jumping into a solution attempt. They consider analogous problems, and try special cases and simpler forms of the original problem in order to gain insight into its solution. They monitor and evaluate their progress and change course if necessary. Older students might, depending on the context of the problem, transform algebraic expressions or change the viewing window on their graphing calculator to get the information they need. Mathematically proficient students can explain correspondences between equations, verbal descriptions, tables, and graphs or draw diagrams of important features and relationships, graph data, and search for regularity or trends. Younger students might rely on using concrete objects or pictures to help conceptualize and solve a problem. Mathematically proficient students check their answers to problems using a different method, and they continually ask themselves, "Does this make sense?" They can understand the approaches of others to solving complex problems and identify correspondences between different approaches.

2 Reason abstractly and quantitatively.

Mathematically proficient students make sense of quantities and their relationships in problem situations. They bring two complementary abilities to bear on problems involving quantitative relationships: the ability to *decontextualize*—to abstract a given situation and represent it symbolically and manipulate the representing symbols as if they have a life of their own, without necessarily attending to their referents—and the ability to *contextualize*, to pause as needed during the manipulation process in order to probe into the referents for the symbols involved. Quantitative reasoning entails habits of creating a coherent representation of the problem at hand; considering the units involved; attending to the meaning of quantities, not just how to compute them; and knowing and flexibly using different properties of operations and objects.

3 Construct viable arguments and critique the reasoning of others.

Mathematically proficient students understand and use stated assumptions, definitions, and previously established results in constructing arguments. They make conjectures and build a logical progression of statements to explore the truth of their conjectures. They are able to analyze situations by breaking them into cases, and can recognize and use counterexamples. They justify their conclusions,

Copyright 2010. National Governors Association Center for Best Practices and Council of Chief State School Officers. All rights reserved.

communicate them to others, and respond to the arguments of others. They reason inductively about data, making plausible arguments that take into account the context from which the data arose. Mathematically proficient students are also able to compare the effectiveness of two plausible arguments, distinguish correct logic or reasoning from that which is flawed, and—if there is a flaw in an argument—explain what it is. Elementary students can construct arguments using concrete referents such as objects, drawings, diagrams, and actions. Such arguments can make sense and be correct, even though they are not generalized or made formal until later grades. Later, students learn to determine domains to which an argument applies. Students at all grades can listen or read the arguments of others, decide whether they make sense, and ask useful questions to clarify or improve the arguments.

4 Model with mathematics.

Mathematically proficient students can apply the mathematics they know to solve problems arising in everyday life, society, and the workplace. In early grades, this might be as simple as writing an addition equation to describe a situation. In middle grades, a student might apply proportional reasoning to plan a school event or analyze a problem in the community. By high school, a student might use geometry to solve a design problem or use a function to describe how one quantity of interest depends on another. Mathematically proficient students who can apply what they know are comfortable making assumptions and approximations to simplify a complicated situation, realizing that these may need revision later. They are able to identify important quantities in a practical situation and map their relationships using such tools as diagrams, two-way tables, graphs, flowcharts and formulas. They can analyze those relationships mathematically to draw conclusions. They routinely interpret their mathematical results in the context of the situation and reflect on whether the results make sense, possibly improving the model if it has not served its purpose.

5 Use appropriate tools strategically.

Mathematically proficient students consider the available tools when solving a mathematical problem. These tools might include pencil and paper, concrete models, a ruler, a protractor, a calculator, a spreadsheet, a computer algebra system, a statistical package, or dynamic geometry software. Proficient students are sufficiently familiar with tools appropriate for their grade or course to make sound decisions about when each of these tools might be helpful, recognizing both the insight to be gained and their limitations. For example, mathematically proficient high school students analyze graphs of functions and solutions generated using a graphing calculator. They detect possible errors by strategically using estimation and other mathematical knowledge. When making mathematical models, they know that technology can enable them to visualize the results of varying assumptions, explore consequences, and compare predictions with data. Mathematically proficient students at various grade levels are able to identify relevant external mathematical resources, such as digital content located on a website, and use them to pose or solve problems. They are able to use technological tools to explore and deepen their understanding of concepts.

6 Attend to precision.

Mathematically proficient students try to communicate precisely to others. They try to use clear definitions in discussion with others and in their own reasoning. They state the meaning of the symbols they choose, including using the equal sign consistently and appropriately. They are careful about specifying units of measure, and labeling axes to clarify the correspondence with quantities in a problem. They calculate accurately and efficiently, express numerical answers with a degree of precision appropriate for the problem context. In the elementary grades, students give carefully formulated explanations to each other. By the time they reach high school they have learned to examine claims and make explicit use of definitions.

7 Look for and make use of structure.
Mathematically proficient students look closely to discern a pattern or structure. Young students, for example, might notice that three and seven more is the same amount as seven and three more, or they may sort a collection of shapes according to how many sides the shapes have. Later, students will see 7 × 8 equals the well remembered 7 × 5 + 7 × 3, in preparation for learning about the distributive property. In the expression $x^2 + 9x + 14$, older students can see the 14 as 2 × 7 and the 9 as 2 + 7. They recognize the significance of an existing line in a geometric figure and can use the strategy of drawing an auxiliary line for solving problems. They also can step back for an overview and shift perspective. They can see complicated things, such as some algebraic expressions, as single objects or as being composed of several objects. For example, they can see $5 - 3(x - y)^2$ as 5 minus a positive number times a square and use that to realize that its value cannot be more than 5 for any real numbers x and y.

8 Look for and express regularity in repeated reasoning.
Mathematically proficient students notice if calculations are repeated, and look both for general methods and for shortcuts. Upper elementary students might notice when dividing 25 by 11 that they are repeating the same calculations over and over again, and conclude they have a repeating decimal. By paying attention to the calculation of slope as they repeatedly check whether points are on the line through (1, 2) with slope 3, middle school students might abstract the equation $(y - 2)/(x - 1) = 3$. Noticing the regularity in the way terms cancel when expanding $(x - 1)(x + 1)$, $(x - 1)(x^2 + x + 1)$, and $(x - 1)(x^3 + x^2 + x + 1)$ might lead them to the general formula for the sum of a geometric series. As they work to solve a problem, mathematically proficient students maintain oversight of the process, while attending to the details. They continually evaluate the reasonableness of their intermediate results.

Connecting the Standards for Mathematical Practice to the Standards for Mathematical Content
The Standards for Mathematical Practice describe ways in which developing student practitioners of the discipline of mathematics increasingly ought to engage with the subject matter as they grow in mathematical maturity and expertise throughout the elementary, middle and high school years. Designers of curricula, assessments, and professional development should all attend to the need to connect the mathematical practices to mathematical content in mathematics instruction.

The Standards for Mathematical Content are a balanced combination of procedure and understanding. Expectations that begin with the word "understand" are often especially good opportunities to connect the practices to the content. Students who lack understanding of a topic may rely on procedures too heavily. Without a flexible base from which to work, they may be less likely to consider analogous problems, represent problems coherently, justify conclusions, apply the mathematics to practical situations, use technology mindfully to work with the mathematics, explain the mathematics accurately to other students, step back for an overview, or deviate from a known procedure to find a shortcut. In short, a lack of understanding effectively prevents a student from engaging in the mathematical practices.

In this respect, those content standards which set an expectation of understanding are potential "points of intersection" between the Standards for Mathematical Content and the Standards for Mathematical Practice. These points of intersection are intended to be weighted toward central and generative concepts in the school mathematics curriculum that most merit the time, resources, innovative energies, and focus necessary to qualitatively improve the curriculum, instruction, assessment, professional development, and student achievement in mathematics.

Copyright 2010. National Governors Association Center for Best Practices and Council of Chief State School Officers. All rights reserved.

Chapter One: Conceptual Foundations

Class Activity 1a: Feet and Hands

How dare we speak of the laws of chance? Is not chance the antithesis of all law?
Bertrand Russell

Each of you should toss a coin one time. If it lands heads, measure the length of your foot in centimeters (either foot, barefoot); and if it lands tails, measure the length of your hand in centimeters (either hand). Then measure your height in centimeters. On the chalkboard, make two columns -- one for height, and the other for foot/hand length. Record your two measurements, without identifying whether you measured your hand or your foot. Then each group should answer the following questions. Afterwards, we will discuss the answers and the methods you used.

1) Which of the measurements are of hands and which are of feet? How do you know?

2) If you saw the print of a bare foot in the mud, left by a criminal, and it measured 25 cm, how tall would you expect the criminal to be? What if it measured 30 cm? Describe any factors that might affect your answer.

Class Activity 1b: Tuition and Minimum Wage

The table below shows the average tuition and fees for 4-year public universities in America, in dollars, for the academic year ending as stated (ref 8), and the federal minimum wage, in dollars per hour (ref 9).

Year	Tuition & Fees	Minimum Wage
1977	617	2.30
1982	909	3.35
1987	1414	3.35
1992	2107	4.25
1997	2975	5.15
2002	3766	5.15
2007	5836	5.85

a) Draw a scatterplot of the variables *tuition & fees* and *minimum wage*, and draw a straight line that you think best represents the linear relationship between these two variables. Find the equation for your line.

b) Use the equation for your line from *part a)* to predict *tuition & fees* in a year when the minimum wage was $3.80. This is called **interpolation**, because you are predicting *within* the range of the given data.

c) Use the equation for your line from *part a)* to predict *tuition & fees* in a year when the minimum wage was 40¢. (Yes, there was such a time!) This is called **extrapolation**, because you are predicting *outside* the range of the given data. Why might extrapolation be an untrustworthy procedure?

d) What is the slope of your line? Interpret the meaning of this number in the context this data (that is, what does this number say about tuition and minimum wage? What is the y-intercept of the regression line? Why does the y-intercept in this example not have a very meaningful interpretation?

e) Since the scatterplot of *tuition & fees* vs. *minimum wage* falls pretty closely along a line, does this indicate that a university education has remained about equally affordable over this time period, in comparison to wages? Conduct some further analysis of the data to investigate this question.

Read and Study: The Language of Statistics

He uses statistics as a drunken man uses lampposts
– for support rather than illumination.
Andrew Lang

It was the year 1618. Johannes Kepler had just described of the motion of the planets around our sun. His result was simple and elegant, but the process had been agonizing. For years Kepler had struggled to establish that the shape of an orbit was one inspired by God; he found that the paths were elliptical, but he did not see divine inspiration in that shape, and so he never properly appreciated his own discoveries (ref 1).

The point we want to make here is that Kepler did not know the exact locations of the planets. He drew his conclusions from measurements recorded by himself and his mentor, Tycho Brahe. Of course these measurements, like all measurements, were imperfect (Tycho even did his without the aid of a telescope), and they were made only at certain points in time.

Hey, pay attention. We have just described the very nature of **statistics**, the study of data collected from a **sample** in order to make inferences about a larger **population** from which that sample was drawn. In this case the population is all the precise positions of the planets at all times (information Kepler didn't have); the sample is the set of approximate positions that Kepler used in his work. By an "inference" we just mean a conclusion about what is true about the population.

When data from a *sample* is summarized with a single representative value -- for instance, the sample's mean, median, standard deviation, etc. (we will study these concepts in more detail later in the book). Such a value is called a **statistic**. Sometimes we will stress that the statistic comes from the sample by using the phrase "sample statistic", but in fact all statistics come from a sample. What we really desire is the corresponding value that represents the entire *population* -- this value is called a **parameter** -- but we don't usually have access to that much information. Examples of a population parameter include a *population mean* or a *population standard deviation.* In short, we use a statistic taken from a sample to estimate or make inferences about a parameter for a population.

In statistics, a variable is some characteristic that can change from person to person, or from observation to observation. In Kepler's case, the variables under study were numeric locations for the planets, and it was possible to make arithmetic computations with them. Variables like that we call **numerical** or **quantitative**. In other studies it might not make sense to do arithmetic with the data. The data might be favorite movies or presidential candidates, that is, the data could be **categorical**. *Give some examples of questions that might request numerical data. Categorical data? Is your Social Security Number a numeric variable? Explain.*

The types of inferences we seek in statistics can vary widely. For example, we might try to describe the "center" of a population with a single number, we might compare some feature of two different populations, or we might investigate the relationship between two variables within a population. This last instance is the type you explored in the *Feet and Hands* class activity. The other types of inference will require an understanding of probability theory; we'll do that later.

When we want to visualize the relationship between two variables (when both of those variables are quantitative) it can be helpful to draw a type of graph called a **scatterplot**. That's a two-dimensional graph, where each data point has two coordinates.

Check out the scatterplot below. It shows the performance of the Dow Jones Industrial Average (DJI) for the month of September, 2001. The DJI is a weighted average of stock prices of 30 major U.S. companies in the New York Stock Exchange, designed to be a reflection of the health of the U.S. economy. One variable in the scatterplot is the date (which in some contexts could be thought of as categorical, but which, in this instance, we regard as quantitative), and the other is the DJI daily closing value. (The data for this chart as well as the two to follow were obtained from Yahoo's finance page (ref 2).)

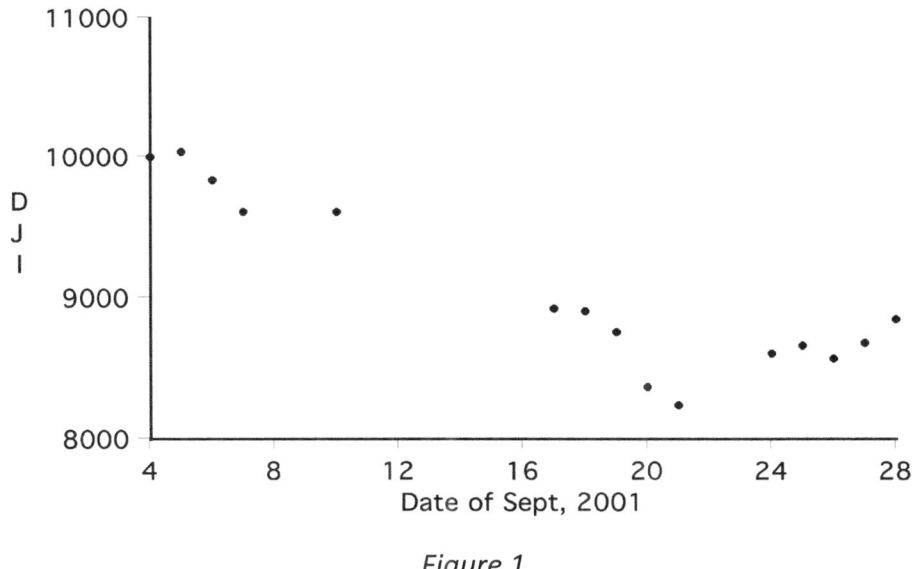

Figure 1.

Take a minute to study the graph. What do you conclude? Describe what happened in September of 2001.

It might appear from the graph that the market really suffered during this time period, but *it is important to understand the scale used on the axes*. To put things into perspective, over the 10 years that preceded this month, the DJI ranged from about 3,000 to about 12,000; we should take this into consideration when deciding how bad this drop really was. The book *How to Lie with Statistics* (ref 3) (which is dated but still useful and easy to read), shows several more examples of graphs that are misleading due to scaling distortions.

Now check out this graph of the Nasdaq Composite Index (which is based on over 3,000 companies in the Nasdaq market).

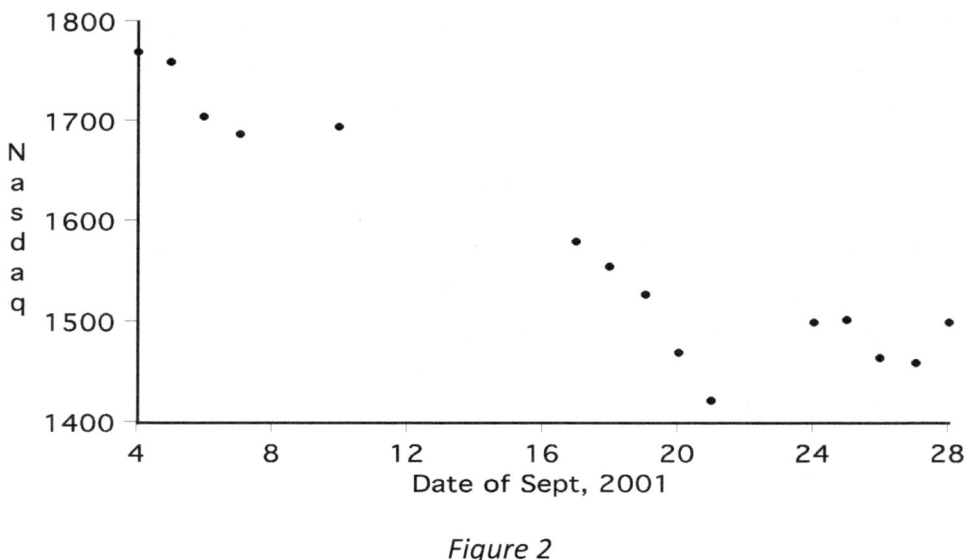

Figure 2

You might have noticed that the two previous scatterplots have a similar shape. Now, rather than graphing the DJI and the Nasdaq separately over time, we'll compare the two indices directly, removing the date variable, with the scatterplot in *Figure 3*. *Make sure that you understand this graph.*

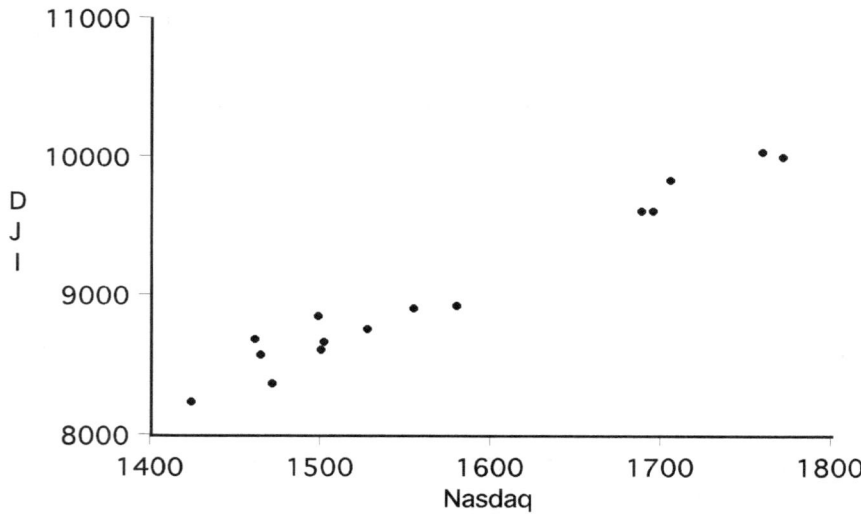

Figure 3.

Notice in this plot that increases in one variable tend to correspond to increases in the other variable; we say these variables have a **positive association**. Further, in this example, the points almost lie along a straight line (with positive slope); so we can make the stronger statement that these two variables have a **positive linear correlation**. (If instead, increases in one variable had corresponded to *decreases* in the other variable, then we would replace the word "positive" with "negative" in those terms.) The more closely the points come to actually lying along a line, the stronger we say the linear correlation is. This idea is formalized through a **correlation coefficient** – you might see it, denoted by r, when working with scatterplots on a computer. We will study the mathematical formulation of this later, but for now, we'll just state a few basic facts about the correlation coefficient:

- It can be computed only when both variables are quantitative (numerical); the concept of correlation does not apply to categorical variables.

- It is always a value between -1 and 1 (negative or positive matching the slope of the line, and the extreme values occurring when all the points lie exactly on a line).

- It only measures the strength of a *linear* correlation; in fact, r might be close to 0 even if the variables have a strong association of some type, as long as that association is not linear.

One should not draw too much information out of even a strong association. The plot in *Figure 3* indicates that both indices were consistent in their reflection of the economy, but it does not indicate that the stocks in one index had any *influence* over stocks in the other. The fact that two variables are related does not imply that they are causally related. When there is an association between two variables, for example when there is a strong linear correlation between two numerical variables, X and Y, this association may be due to one or more of the following phenomena:

- **Cause and effect relationship:** variable X causes variable Y (or the other way around). In other words, the values taken by variable Y change because the values taken by variable X change. (This is the idea behind the terms *explanatory* and *response* variables).

- **Confounded relationship:** X and Y are associated because there is a confounding variable Z that is associated with both X and Y, so that we cannot distinguish between the effect of X on Y and the effect of Z on Y. A common confounding scenario is where there is cause and effect relationship between Z and X, and also between Z and Y. The biggest worry in any statistical study is that there could be a confounding variable that we aren't even aware of.

For example, consider one's education level (X) and income (Y). There is a strong positive association between education level and income: the higher the education level (the more years of school attended) the higher the income level.

This could be because there is a cause and effect relationship: the more education you get, the better (the more high-paying) the job you can get. But actually, the cause and effect could go in the other direction: maybe having a higher income means you can afford more education. Or maybe education level and income level are both having a common response to a third variable, such as intelligence. Perhaps smarter people are more likely to get more education, and smarter people are also more likely to make more money. Other lurking variables such as socio-economic class, and personal ambition level could also be associated with both education level and income level, so there may be many confounding variables that could explain the observed relationship between education and income. A challenge for every statistical study is to attempt to account for all potential confounding variables.

Regardless of these ambiguities, statistical tools such as scatterplots can help us to see patterns and provide us with insights. On a scatterplot, if we have data that appears fairly linear, it is often helpful to create a "best" line through the data and then to use that line to make predictions. Typically in statistics, the best line through data is defined to be the one that minimizes the sums of the squares of the vertical distances of the data points to the line. We call this the **least squares regression line**. In practice, we find this line using a calculator or a computer; your instructor can show you how to do this in class using whatever technology you have available. We'll explore the least squares regression line in the homework.

Homework

But in this world nothing can be said to be certain, except death and taxes.
Benjamin Franklin

1) Do all the italicized things in the *Read and Study* section.

2) The data for this problem are for several 2005-model-year sedans in three classes (small, family, and luxury), as determined by Consumer Reports (ref 5). The different classes would usually be regarded as different populations.

Make	Price (dollars)	Weight (pounds)	Stopping Distance from 60 mph (feet)
Class: Small Sedans			
Chevrolet Cobalt	17350	2850	141
Ford Focus	19080	2800	128
Honda Civic	18825	2645	129
Hyundai Elantra	17589	2980	133
Saturn Ion	18415	2865	130
Class: Family Sedans			
Chevrolet Malibu	23935	3535	145
Mitsubishi Galant	27094	3715	144
Nissan Altima	23730	3235	144

Pontiac G6	23080	3475	146
Toyota Camry	21574	3285	147

Class: Luxury Sedans

Acura RL		49670	4035	131
Audi A6		50820	4115	129
Cadillac STS	50335	4030	131	
Infinity M35	50240	4095	122	
Lexus GS300	51859	3915	133	

a) In the table, which of the data is categorical and which is quantitative?

b) Based on the data for *small sedans* and *family sedans* only, what conjecture do you make about the stopping distance compared to the weight of a car? Draw an appropriate scatterplot. Now also include the data for *luxury sedans* in your scatterplot. How does this new data conform to your previous conjecture? Do you want to modify your conjecture?

3) Make up examples (and explain your reasoning in each case) of two numerical variables that you would expect to have a strong **negative** correlation because:
   a) there is a cause and effect relationship.
   b) there is a confounding variable. Explain how this variable could be confounding.

4) Make up a new example (and explain your reasoning in each case) of two numerical variables that you would expect to have a strong **positive** correlation because:
   a) there is a cause and effect relationship.
   b) there is a confounding variable. Explain how this variable could be confounding.

5) The following table depicts the U.S. tax dollars collected by the I.R.S. and the number of deaths in the U.S. for each of five years between 1980 and 2000 (ref 6, 7).

Year	Taxes (in billions of $)	Deaths (in thousands)
1980	519	1990
1985	743	2086
1990	1056	2148
1995	1376	2312
2000	2097	2403

   a) Describe the relationship between these two variables over the years.
   b) My dad thinks that Americans are increasingly being taxed to death. Does the data support this? If so, why? If not, offer an alternative explanation for the connection.
   c) Based on the data, estimate both the number of deaths and the taxes collected for the year 1960. Does your answer seem plausible? Explain.

6) Use a computer spreadsheet program to draw a scatterplot of the variables *tuition & fees* and *minimum wage* in class activity 1b. Use the computer to draw and find the equation of the least squares regression line comparing those two variables. Print out the picture. Now recall that in the reading, we said this line is the one that minimizes the sums *of the squares*

of the vertical distances of the data points to the line. For the line in your picture, draw these squares. (We are asking you to interpret the arithmetic in a geometric way.) Now, to get the idea of how the line was chosen, draw some other line on top of your scatterplot, and then draw the squares corresponding to that line. The total area of the squares for the line your computer drew will be less than the total area of the squares for the line you drew. This is why the line the computer drew is called the ***least squares* regression line**.

7) Design and conduct a study in which you use a scatterplot to investigate whether one minute of vigorous exercise results in a temporarily increased heart rate. (Really do it!)
   a) How many people did you include in your study? How were they chosen? What sort of population did you intend for them to represent?
   b) What type of data did you collect, how did you collect it, and how did you use a scatterplot to graph it?
   c) Did the exercise elevate pulse rates among the people in your study? Explain.
   d) Based on your results, are you willing to conclude that vigorous exercise increases heart rate in humans? Why or why not.

8) Read the Common Core State Standards 8.SP.A.1, 8.SP.A.2 and 8.SP.A.3. Identify exactly which parts of these standards have been addressed so far in this book. Specify an example or activity (from the Class Activity, Read and Study, or Homework) that addresses that competency. Also identify any parts of these standards have not (yet) been addressed in this book.

9) Explore some textbooks or curriculum materials for Grades 7 and 8. Make a copy of two or three activities involving scatter plots and linear associations. Do these activities and write an explanation of how they meet the CCSS standards. Share and discuss these with some of your classmates.

Class Activity 2: HIV and AZT

AZT was the first antiretroviral drug to be used by pregnant women with HIV. In a clinical study published in 1994 by the New England Journal of Medicine, 477 pregnant women with HIV were enrolled in the study and randomly assigned to either a treatment group or control group. The treatment group received the drug AZT during their pregnancy and during childbirth, while the control group received a placebo. Then after the child was born, the infant was tested for HIV. The data yielded 363 mother-infant pairs in which the HIV infection status of the infant was able to be determined. Among these mother-infant pairs, 180 had been assigned to receive AZT and 183 had been assigned to the control group. Thirteen infants in the AZT group were HIV-infected, while in the placebo group, 40 infants were HIV-infected.
(Source: https://www.nejm.org/doi/full/10.1056/nejm199411033311801)

Brainstorm several explanations for why the treatment group has so many fewer HIV positive babies.

Based on this study, do you feel that can we conclude that AZT is effective in reducing HIV transmission from mother to baby? Discuss how reliable you think this evidence is, and why.

Read and Study: Relationships between Categorical Variables

Let's return to the story of statistics. In the previous section, we talked about the typical statistical study. There is some population that we are interested in and some parameter about that population that we would like to know. Perhaps we want to know the median salary of mathematics teachers in Wisconsin, or the correlation between a person's hand length and their height, the proportion of likely voters who will vote for a given candidate, or the percent of times that a 5 turns up when rolling a given die. *In each of these examples we just gave, take a moment to identify the population, and identify the parameter.*

To try to answer our question, we collect data. Rarely do we have data on an entire population. Often we don't even know the size of our population, or our population is a theoretical one. In the class activity, the population of interest was pregnant HIV positive women. Not just the ones in the study, but really all current, future and potential pregnant HIV positive women.

So the data we collect is almost always from some sample that we take from our population. Perhaps we collect data from our class, or survey 500 likely voters, or ask for the salary data from several school districts. Our next task is to analyze that sample data. We make graphic displays of the data, such as a chart, histogram, bar graph, box plot, or line plot. We calculate statistics, such as a frequency, percent, minimum, maximum, mean, medians, standard deviation, etc. When we do these things we are doing **descriptive statistics**, that is, we are describing and summarizing the data in a sample.

These statistics and graphic displays, since they come from a sample, are only estimates of the corresponding parameters and theoretical distribution for the entire population. When we make decisions and predictions about what entire population is like, based on our one sample, we are engaging in **inferential statistics.** In other words, we are using statistics from a sample to make inferences about the entire population. Does AZT reduce the incidence of HIV in babies? What is the relationship between minimum wage and college tuition? Who will win the presidential election? Is this a fair die?

To have confidence that our decisions and predictions will be good ones, we need to know how reliable our data and statistics from our sample really are. So we need to think about all of the potential reasons why our statistics may be unreliable and thus lead to bad decisions or predictions. First we need to worry about the quality of our data, whether we can trust that the data in our sample is representative of our population. We can't tell just by looking at the data. Remember we don't know what the population is like, we only know what our sample is like. So instead we need to consider whether there might be any bias in the *procedures* that generated the data. In general, we say a procedure is **biased** if it systematically favors certain outcomes. A common source of bias is **sample bias,** when the sampling procedures used tend to over or under-represent certain segments of the population being studied.

For example, one of the authors of this text remembers that during the 2008 NFL playoffs, the cell phone company Nextel sponsored a poll where Nextel network callers could call into the CBS NFL broadcast and vote for which quarterback they thought was the best quarterback of all time. Tom Brady won, followed by Peyton Manning, and Dan Fouts. Do these results surprise you? Perhaps not if you aren't an NFL football fan. *Regardless, take a moment now and think about how the sample in this study could be biased.* At the time of the poll, Tom Brady and Peyton Manning were both star players and arguably the top two quarterbacks in the league. But Dan Fouts? A great quarterback, sure, and a Hall-of-famer. But what about Joe Montana? John Elway? Brett Favre? The choice of Dan Fouts is puzzling, until you find out that the poll was conducted during the AFC championship game between the New England Patriots and the San Diego Chargers. Tom Brady was playing in that game for New England and Dan Fouts had played his entire career for San Diego. So while the AFC championship game draws a large general audience of football fans (the intended population of interest), the audience was likely comprised of mostly New England fans or San Diego fans, and so this sample procedure tended to over-represent these segments of the population.

Another source of bias is that only Nextel customers were included in the sample, but it's not obvious that this bias would have a large influence on the results. Sample bias often comes from the subjects self-selecting for the study or experiment, since it's likely that people who volunteer to participate in a certain study have different characteristics than those who don't want to participate. In the above example, it's likely that those fans who feel strongly enough to want to call in (and perhaps accrue charges on their phone) have different opinions from those who didn't call in. These types of "non-scientific") call-in or text-in polls may be entertaining, and may be a great advertising tool for the sponsor, but we need to take their results with a grain of salt.

One word of caution with this example: the idea that people would think that the current star players Brady and Manning were the best all time (neglecting all of the great players in the past) is a form of bias (we can say that people might be biased towards what's current rather than what's old), but this does not mean that this *study* is biased. The study wanted to know what people think, and that's what people think.

Back to why the statistics from our sample may turn out to be unreliable and lead to poor decisions or predictions. As we discussed in the previous Read and Study section, when we are looking for relationships between two variables, we need to be concerned about potential confounding variables. In that last section, we looked only at examples where we had two numerical variables. Confounding can also be a big problem when looking at categorical variables as well. Often our explanatory variable X is some categorical variable that says which of two groups an observation is in: for example, are they Male or Female, or, did they get the treatment drug or a placebo. In these cases, a confounding variable is some third variable Z, other than the category (X) – which affects the responses (Y) being studied, and this variable Z has different levels across the categories of X.

For example, consider the following study, reported in the Journal of Pediatrics in 2007.

To test the association of media exposure with language development in children under age 2 years, the researchers had a sample of 1008 parents of children age 2 months to 24 months who were surveyed by telephone in February 2006. Questions were asked about child and parent demographics, child-parent interactions, and child's viewing of several types of television shows and DVDs/videos. Parents were also asked to complete the short form of the MacArthur-Bates Communicative Development Inventory (CDI). Among infants (age 8 to 16 months), each hour per day of viewing baby DVDs/videos was associated with a 16.99-point decrement in CDI score.

Was it really watching those baby DVDs that caused the infants in the study to have lower language skills? Or could it be that infants with lower language skills watch more baby DVDs (that is, that the cause-and-effect goes the other direction)? Or could there be a confounding variable such that infants with some certain characteristic tend to watch more baby DVDs and also tend to have lower language skills? *Think of some potential confounding variables that could explain the results of this study.*

The authors of the study were very careful not to conclude that watching baby DVDs results in lower language skills, and say that "further research is required to determine the reasons for an association between early viewing of baby DVDs/videos and poor language development."

So far we have discussed sample bias and confounding variables. The best tool we have for dealing with both of these potential issues is random sampling and conducting experiments with assignment. A simple random sampling procedure is unbiased since it does not systematically favor any segments of the population. And randomly assigning our participants to the experimental and control groups is the best way of having both groups have similar distributions of any potential confounding variables.

Lastly, let's consider the following scenario: The National Center on Addiction and Substance Abuse reports that in 2005, 12% of full-time college students in the US smoke daily (source: http://www.casacolumbia.org/templates/publications_reports.aspx). Suppose UW Oshkosh conducts a random survey of 50 UW Oshkosh students and finds that only 3 of the surveyed students smoke daily. Let's suppose that the researchers did a good job with their random sampling procedures and that there is no sample bias or confounding variables in the study. Our fundamental question remains, how good of an estimate is the sample statistics (6% of the sample smokes) for the population parameter (the percent of all UW Oshkosh students that smoke). Does this result really indicate that the percent of UW Oshkosh students who smoke daily is significantly less than the national rate (12%)? Maybe just by random chance we happened to get a sample that had few smokers. To assess this, we really need to know how likely it is that we could sample results like this given different possibilities for the population parameter. To do this, we need to know a bit about probability. A major goal of this text is to learn enough about probability to be able to do just these types of assessments and be able to say just how reliable our sample results really are.

Homework

1) In the spring of 1991, Major League Baseball conducted a survey of alcohol and drug use among its players. Of the 880 players who responded (representing about 80% of major league players at that time), only 1.5% reported using anabolic steroids during their lifetime, and only 0.5% reported use of steroids in the preceding year. (Source: http://mlb.mlb.com/mlb/news/mitchell/report.jsp)Discuss some likely sources of sample bias in this study and how they may have influenced the results.

2) The most extensive public opinion poll in history was conducted by Literary Digest magazine for the 1936 presidential election. This magazine had correctly predicted the results of the previous 5 elections. This time, they bragged that they will ask a quarter of the countries voters who they will be voting for. They mailed out their survey to over 10 million people, getting names and addresses from their magazine's subscription list, phone books, vehicle registration lists and club memberships. More than 2.4 million people returned postcard survey, with 57% saying that they would be voting for Republican Alf Landon. But Democrat Franklin Roosevelt ended up winning re-election, with a whopping 63% of the vote. (Source: http://historymatters.gmu.edu/d/5168/ What went wrong? Try to think of some sources of sample bias in this study.

3) Suppose you conduct a survey to study the effectiveness of vitamins on preventing the flu, and you survey a random sample of 120 college students. You ask the students whether they take a daily multi-vitamin, and whether they got the flu in the past year. The hypothetical results are shown below.

	Vit.	Flu		Vit.	Flu		Vit.	Flu		Vit.	Flu
1	N	N	31	N	Y	61	Y	N	91	N	Y
2	N	N	32	Y	N	62	N	N	92	N	Y
3	Y	Y	33	Y	N	63	Y	N	93	N	Y
4	N	N	34	N	Y	64	N	N	94	N	N
5	Y	N	35	Y	N	65	Y	Y	95	Y	Y
6	N	N	36	Y	N	66	Y	N	96	Y	N
7	Y	Y	37	Y	N	67	Y	N	97	N	N
8	N	N	38	Y	Y	68	N	Y	98	N	Y
9	N	N	39	N	N	69	Y	N	99	Y	N
10	Y	N	40	Y	Y	70	Y	N	100	N	Y
11	Y	N	41	N	N	71	Y	N	101	N	N
12	N	Y	42	Y	Y	72	N	N	102	Y	Y
13	N	N	43	N	N	73	Y	N	103	N	Y
14	N	Y	44	N	N	74	Y	N	104	N	N
15	Y	N	45	Y	N	75	N	Y	105	N	N
16	Y	N	46	N	Y	76	N	N	106	N	N
17	N	N	47	N	N	77	N	N	107	Y	N

18	N	N	48	N	N	78	N	Y	108	Y	N
19	Y	N	49	N	N	79	N	N	109	N	N
20	N	Y	50	Y	N	80	N	Y	110	Y	N
21	Y	N	51	N	Y	81	N	N	111	Y	N
22	N	Y	52	N	N	82	Y	N	112	Y	Y
23	N	N	53	N	N	83	Y	N	113	N	Y
24	N	N	54	N	N	84	Y	N	114	Y	Y
25	N	N	55	N	N	85	N	Y	115	N	N
26	N	Y	56	N	Y	86	N	N	116	Y	N
27	Y	N	57	N	N	87	N	N	117	N	N
28	N	N	58	Y	N	88	N	N	118	Y	Y
29	N	Y	59	N	N	89	Y	N	119	Y	Y
30	N	Y	60	N	N	90	N	N	120	N	N

Source: completely made up by one of the authors).

a) Based on the results of this study, is this vaccine effective in preventing the flu? Justify your response.
b) Report as many percentage statistics as you can from this data. Which statistics are misleading, and why? Which statistics are most appropriate in describing the relationship between getting the vaccine and getting the flu?
c) A confounding or lurking variable is an extraneous variable in a statistical model that may have caused the observed effect. Discuss any likely confounding variables in this study.
d) Think of some ways in which the study could suffer from sample bias.
e) Do you think the results of the study could be due to random chance? Explain.

4) Now back to some actual data. The following table reports the new admissions to Wisconsin prisons in 1999. Source: http://www.ssc.wisc.edu/~oliver/RACIAL/DOCdata2001/WiscPrisonAdmissions.htm. Describe several relationships you can find between prison admissions and race in these data. Use relevant calculations to support your claims.

	White	Black	Hispanic	Am.Indian	Asian
VIOLENT OFFENSES	390	280	63	26	13
ROBBERY/BURGLARY	157	208	25	11	5
DRUG OFFENSES	139	539	79	5	2
LARCENY/THEFT	129	122	11	4	2
OTHER OFFENSES	195	113	14	25	1
UNKNOWN	11	4	0	0	0
Total	1,021	1,266	192	71	23
WI Population	4,701,123	285,308	140,235	43,534	80,246

5) A 2004 public-health survey in Sweden investigated differences in daily smoking prevalence in different age groups. (Source: Scandinavian Journal of Public Health, 2009; 37: 146–152). The researchers mailed a questionnaire to a random sample of approximately 47,000 adults in southern Sweden. A total of 27,757 persons aged 18–80

years responded. A key result was that 14.9% of the men and 18.1% of the women were daily smokers. The authors of the study recommended strategic public policy interventions to reduce women's smoking.
   a) Explain how sample bias may have influenced the results of the study.
   b) The authors looked at several potential confounding variables in their study that could be related to this association between gender and daily smoking. Brainstorm a couple of potential confounding variables and describe how they might explain the key result of the study.
   c) Explain how random chance may have influenced the results of the study.

6) Make up some new examples (and explain your reasoning in each case) of two categorical variables that you would expect to have a strong association because:
   a) there is a cause and effect relationship.
   b) there is a confounding variable.

7) At a certain University, there are 122 math majors. 89 are male and 33 are female. The math majors need to choose an emphasis: Education, Statistics, Applied, or Liberal arts. 82 of the majors have chosen an Education emphasis. Of these 61 are male and 21 are female. Does this data suggest that there is relationship between a math major's Gender and whether they choose the Education Emphasis? Explain.

8) Suppose you conduct a survey in a small town about hunting and fishing. You asked 100 respondents whether they liked to fish (Yes or No) and whether they liked to hunt (Yes or No). Suppose 80 of the respondents said "Yes" to whether they liked to fish, and 40 of the respondents said "Yes" to whether they liked to hunt.
   a) Decide on a reasonable number of respondents that said "Yes" to both questions so that you could conclude there is no statistical relationship between hunting and fishing in your data.
   b) Decide on a reasonable number of respondents that said "Yes" to both questions so that you could conclude there is a strong statistical relationship between hunting and fishing in your data.

9) Read the Common Core State Standard 8.SP.A.3. Identify exactly which parts of this standards have been addressed so far in this book. For each topic of competency mentioned, specify an example or activity (from the Class Activity, Read and Study, or Homework) that addresses it. Also identify any parts of this standards have not (yet) been addressed in this book.

Class Activity 3: It's a Long Shot

The excitement that a gambler feels when making a bet is equal to the amount he might win times the probability of winning it.
Blaise Pascal

The main event for today is a *Long Shot*. Someone from the class will be chosen to attempt to shoot a wadded piece of paper across the room into a wastepaper basket (without any practice attempts).

Our question of interest is this: *What is the probability that the person chosen makes the shot?*

Thinking ahead:
Before we do anything and before you read ahead…
Discuss in your group what you might do to answer this question (remember that the person does *not* get *any* practice throws).

Preparing for the Main Event:
We will now prepare for the main event as follows:
1. Your teacher will select the person that will attempt the shot.
2. You will each receive one "dollar" for betting purposes.
3. On the back of the "dollar", write whether you think the person will make or miss the shot.
4. Now each of you will now (in turn) tell whether you are going to bet your "dollar" on the person making the shot or missing the shot. Decide carefully because once you have placed your bet, you cannot change it.
5. With your group, determine the answer to our big question:
What is the probability that the person chosen makes the shot?

The Main Event:
One more thing before the main event…
What is the probability that the person chosen will make the shot?
Justify your decision.

Now for the shot…

Read and Study: Concepts of Probability

Life is a school of probability.
Walter Bagehot

When we talk about the **probability** of something happening, we mean to quantify how *likely* it is to happen. Of course if one person was using a 1-to-100 rating scale and her friend was using a 4-star system, they might have a hard time communicating, so the mathematical community has agreed to use a 0-to-1 scale for probabilities. If something could never happen its probability is 0. Most people would agree that the probability of the moon's dropping out of the sky tomorrow and crashing to earth is pretty close to 0. If something will definitely happen, its probability is 1. Most people in Wisconsin (but maybe not Alaska) would agree that the probability of the sun rising tomorrow morning is pretty close to 1. After that, things get more complicated. If you toss a coin, for example, what is the probability of it landing heads? If it is well balanced and symmetric, you might say ½ without even bothering to toss it. Alternatively, if you toss it many times and discover that it lands heads half the time, you might also say ½. But under this second interpretation, several issues arise; we will devote the next section to exploring these issues.

You may be surprised to hear that mathematicians cannot agree on exactly what we mean when we speak of the probability of something; some contrasting views and their histories are presented in (ref 1). Part of the disagreement, as already indicated, regards how we should evaluate the probability of an event (that is, what number should be chosen, and why); the other part of the disagreement regards how the probabilities of various events should interact with each other, mathematically.

It is generally agreed upon that a probability model must begin with some sort of trial in mind, whose set of possible outcomes (called a **sample space**) is specified, and that probabilities can be defined for various collections of these outcomes (called **events**). For example, suppose you are rolling a six-sided die. Your sample space might consist of the individual outcomes *1, 2, 3, 4, 5,* and *6*. The event *rolling an even number* would then consist of the outcomes *2, 4,* and *6*. Most (but not all) mathematicians agree that things should "add up right," in the following sense: the probabilities of two **mutually exclusive** events (events which share no common outcomes), when added together, should equal the probability of the *union* of those events. In our die example, the probability of rolling a *2* plus the probability of rolling a *4* plus the probability of rolling a *6* must equal the probability of rolling an even number. The mathematical requirements beyond this stage are a topic of more heated debate.

Regardless of our disagreements, for all of us, probability reflects a *degree of belief*, and this is a strong enough common ground to serve many purposes. You will encounter a variety of settings as you read this book. For now, we will mention only that gambling has played a central role in the development of probability theory, from its origins in the 1600's (ref 2) to its widespread use in the modern stock market (ref 3). That's why we introduced the topic through gambling in *Activity 2*, and also in some of the *Homework*. In this context, a probability is a price you would be willing to pay for a $1-payout bet, if everything is fair. *Stop and think about this*.

A note regarding terminology: In the context of gambling, people often describe probabilities in terms of **odds**. In Activity 2, if twice as much money were being placed on making the shot as on missing it, then the house would post the odds of *making the shot* as 2:1 (we say "two-to-one"), since for every $2 placed on *making the shot*, only $1 is placed on *missing the shot*. The probability associated with these odds is $\frac{2}{3}$ in favor of making the shot, since $\frac{2}{3}$ of the money is being placed on making it. The implied assumption is that $\frac{2}{3}$ of the "belief of the bettors" is on the shooter's side.

We will make a distinction between **theoretical** probability and **experimental** probability. The **experimental probability** of an event is calculated by actually conducting the random experiment repeatedly (many trials) and observing the frequency at which the event occurs. We can assign experimental probabilities to events really only when our trial is easy to replicate. For example, if our trial is rolling a die, we can easily repeat this many times and assign an experimental probability to rolling, say a 5, as the proportion of all of your rolls that resulted in a 5.

But before rolling the die even once, it makes sense to assign a probability of 1/6 to getting a five because we expect that each of the six sides of the die is equally likely to end face up. This is called a **theoretical** probability, as it is based on what we know about the random experiment prior to any actual trials. In cases where it is possible to conduct the randomly experiment repeatedly, we expect that in the long run, the experimental probability will match closely to the theoretical probability. However, in many situations, the experiment is not repeatable, such as in a horse race, or a presidential race, or a rocket launch. We really want to know the probability of someone winning (or the probability of something exploding) beforehand. This is where theoretical probability comes in. And as the class activity illustrates, the collective belief that an event is a theoretical probability and one way to quantify this belief is the price one would be willing to pay to win $1.

Connections to Teaching: Quantifying Chance

Teachers should give middle-grades students numerous opportunities to engage in probabilistic thinking about simple situations from which students can develop notions of chance.
NCTM Principles and Standards

Middle grades teachers have the job of helping students to quantify chance, that is, to assign probabilities to events. In elementary grades children will have discussed events as impossible, unlikely, likely or certain. They will have performed simple experiments -- tossing coins, dice (they call these 'number cubes' in schools), or even marshmallows -- to compute experimental probabilities as fractions. They will have graphed the results of their experiments using pie charts, bar graphs and pictograms. You will build from these experiences

The *Connected Mathematics Project* curriculum for Grade 6 supports the notion of probability as a 'degree of belief'. In their unit titled *How Likely is It?*, students are asked to first, give an fraction or percentage representing how likely they feel certain things have of happening; and then, justify the number they chose. As an example, it states:

> "I watch some television every night unless I have too much homework. So far today I do not have much homework. Therefore, I am about 95% sure that I will watch television tonight." (p.?)

Following this example, give an estimate (as decimal, fraction, or percent) for the probability of each of the following happening. Be sure to justify your answer by explaining your reasoning: (items are variations of those provided in the aforementioned curriculum)

1) You will miss at least one day of class this semester.
2) The sun will set tonight.
3) You will get at least 40 emails this week.
4) You will have cereal for breakfast at least one day this week.
5) We will not get any snow in Wisconsin this year.
6) The next person to enter the classroom door is a girl.

How would this activity help students' understanding of probability?

NCTM's *Illuminations* activities online include three activities in which students draw upon experimental & theoretical probabilities to understand fair games (*Is It Fair, A Fair Hopper*, and *Happy Hopper*). Examine them at https://illuminations.nctm.org/Lesson.aspx?id=1145. What can students learn about probability & fairness from each one? How does each build upon students' understanding of probability?

Homework

> *"I think you're begging the question," said Haydock, "and I can see looming ahead one of those terrible exercises in probability where six men have white hats and six men have black hats and you have to work it out by mathematics how likely it is that the hats will get mixed up and in what proportion. If you start thinking about things like that, you would go round the bend. Let me assure you of that!"*
> *Agatha Christie*

1) Do all those problems and middle school activities given in the *Connections* section. Bring them for discussion.

2) A commonly used formula to compute probabilities is that the probability of an event is the number of outcomes in that event, divided by the number of outcomes in the sample space, which could be written $P(E) = n(E)/n(S)$. In this problem you will think about when this formula will and won't work to find probabilities of events. You are presented with a random experiment and a sample space, and an event, and a probability for that event. For each scenario, identify the sample space S, the event E, the number of outcomes in the sample space $n(S)$ and the number of outcomes in the event $n(E)$. In which case(s) is reasoning correct? If the stated probability is not correct, what is the flaw in the reasoning, and what is the correct probability?
    a) Consider the random experiment of flipping a coin twice. The result could be that you get 0 heads, 1 heads, or 2 heads. So the probability of getting no heads is 1/3.
    b) Consider again the random experiment of flipping a coin twice. The result could be that the coins match, or the coins are different. So the probability that the coins match is ½.
    c) Consider the random experiment of selecting an M&M from a new one-pound bag of regular M&M's. Since there are 6 colors of M&M's (red, yellow, blue, green, orange, and brown), the probability that you select a yellow one is 1/6.
    d) Two standard six-sided dice are rolled and the sum of the two numbers are recorded. The sum could be any number from 2 to 12, so the probability that you roll a sum of 5 is 1/11.
    e) Under what conditions is the formula $P(E) = n(E)/n(S)$ valid?

3) Alfred Hitchcock's movie *Rear Window* is 1 hour and 54 minutes long. Hitchcock himself makes a 10-second appearance, beginning 26 minutes into the movie. On Saturday, the movie cinema *The River Oaks* will begin showing this movie (with no previews) at precisely noon. They will continue to show it every 2 hours (with 6-minute breaks between showings), around the clock until further notice. On Friday, you loosen the hour and minute hands on your clock, and spin them, to generate a random time. On Saturday, you will go to *The River Oaks* and enter the theater at exactly that randomly-chosen time. (Work with us here.)

    a) What is the probability that Hitchcock will be on the screen the moment you enter the theater?
    b) Upon entering the theater, you remain until the end of that showing (or the one that is just about to start). What is the probability that you will see the entire Hitchcock scene?

4) A man works the late shift in Manhattan, and then rides the southbound subway train home to Brooklyn. Because his work schedule is erratic, he always arrives at his subway station (57th St. & 7th Av.) at some random time between midnight and 5 AM. He

notices that about 2/3 of the time, the northbound train arrives before his southbound train. But in fact the schedule confirms that the northbound trains arrive like clockwork every 20 minutes, and so do the southbound trains. How could this be?

5) Suppose you drop a ball whose diameter is 1 inch through a square grid of laser beams spaced 2 inches apart, creating 2 inch by 2 inch square gaps. What is the probability that the ball will pass through the grid without hitting any of the laser beams? (Assume that this is a large grid, and you are nowhere near the boundaries.)

6) Here is one of the earliest known problems in the history of probability theory. Gamblers playing dice around the year 1600 wanted to know: If three dice are rolled, which is more likely, to roll a sum of 9, or to roll a sum of 10, or are both equally likely? Investigate this question and see if you can determine and justify an answer.

7) There is a 25% chance of rain on Saturday and a 50% chance of rain on Sunday. Assess the probability that it will rain this weekend.

8) Three men are given hats to wear. Each hat color is randomly chosen to be either red or blue. Every man can see the hats on the heads of the other two men, but not the hat on his own head. At a given point in time, the men will be taken to separate rooms, and each asked to guess the color of his own hat. Each man may choose whether or not to answer the question. If anybody answers incorrectly, or if nobody answers, then the men lose the game. But if at least one person answers correctly and nobody answers incorrectly, then everybody wins a million dollars. The men may discuss a strategy before the game begins, but they are not allowed to communicate with each other after the hats have been placed on their heads. What is the best strategy, and how likely are they to win the million dollars using this strategy?

9) You and your friend, nicknamed "Gullible," are preparing to place bets with each other on the outcome of a horse race. Five horses (A, B, C, D, and E) will be racing. You have allowed Gullible to stipulate the options for the betting, and, in exchange, you get to decide which bets are placed. Gullible has decided that you can bet only on the following blocks of horses, and at the following prices:

Block of Horses	Price for a 1-dollar payoff bet
A & B & C	60¢
D & E	50¢

So, for example, you cannot bet on the horse A by itself, because that option isn't listed. But for 60¢, you could buy a bet on the block of horses A&B&C together, so that if any one of them wins, then Gullible will have to pay you 1 dollar, and you will make a 40¢ profit. Here's another example: Suppose you bought a bet on A&B&C, and you sold two bets on D&E. If A or B or C wins the race, then you win a 1-dollar payout, minus the 60¢ you paid for your bet, plus the dollar for the two bets you sold, thereby earning a net profit of $1.40. However, if D or E wins, then you are hit with a net loss of $1.60 (you have lost two dollars, and you still have to figure in the price of the bets).

a) It turns out that Gullible chose his prices badly. Find a combination of bets by which you are guaranteed, regardless of which horse wins the race, to earn a positive net profit.

b) Suppose instead that Gullible has stipulated the following blocks of horses and prices (but that he has not yet decided on a price for C):

Block of Horses	Price for a 1-dollar payoff bet
A & B & C	40¢
C & D & E	75¢
C	?

In order to avoid a situation such as in part a), where you could guarantee yourself a winning bet, how much should he charge for a bet on C?

10) Read the Common Core State Standards for Grade 7 under: **Investigate chance processes and develop, use, and evaluate probability models** (7.SP.C.5-7.SP.C.8) Identify exactly which parts of these standards have been addressed so far in this book. For each topic of competency mentioned, specify an example or activity (from the Class Activity, Read and Study, or Homework) that addresses it. Also identify which parts of these standards have not (yet) been addressed in this book.

Class Activity 4: Spaghetti Triangles

Lest men suspect your tale untrue, keep probability in view.
John Gay

To prepare for this activity, **each** of you will need to get:
- Twenty unbroken pieces of spaghetti (usually about 25 cm long)
- A way of generating random numbers uniformly from, say, 0.1 to 24.9
- A ruler with centimeters.

Once you have your materials, **each** of you should complete *Task A* and record your data from at the board, before moving on to *Task B*.

Task A: *Each of you should do the following 10 times…*
- Take an unbroken spaghetti noodle and randomly (using your intuitive feel for randomness) break it into 3 pieces and try to make a triangle out of those three pieces.
- Record the length of the largest piece and whether or not you were able to make a triangle.

Task B: *Each of you should do the following 10 times…*
- Use a random number generator to generate two random numbers between 0 and 250.
- Measuring (in mm) an unbroken spaghetti noodle from one end, mark two spots on the noodle corresponding to the two random numbers you generated.
- Break the noodle at those two locations—so again, you'll end up with 3 pieces—and try to make a triangle out of those three pieces.
- Record the length of the largest piece and whether or not you were able to make a triangle.

Once you have completed these tasks, continue onto the next page.

Questions for discussion:
- How can you use the data from each task to the following question: (Actually do it once the data is available.)

    If a spaghetti noodle were to randomly break into three pieces, what is the probability that those three pieces could form a triangle?

- Do your results from Task B match the ones from Task A? If not, why do you think this is so?

- How could you calculate this probability theoretically? Justify your approach.

Read and Study: Randomness

To us probability is the very guide of life.
Bishop Butler

Earlier we suggested that you might evaluate the probability of a coin landing heads by tossing it many times. If it lands heads half the time, for example, then you might say its probability of landing heads on a given toss is ½. (This would be classified as an "experimental probability".) But now we must confront some problems.

- How many times is "many"?

- Does "half the time" mean *exactly* half the time? (If so, what would happen if you tossed it just one more time?)

- Does it matter on *which* tosses the heads occur?

- What if you started over and got a different result?

Hmmm... seems problematic. Still repeated trials provide an invaluable means of assigning probabilities. Most of the serious difficulties are resolved by making sure that the number of trials is large enough. (Incidentally, although a large amount of data indicates that coins *do* land heads approximately half the time when tossed, they land tails more often when spun, and heads more often when "tilted." *Try it*. You can read more about all this in (ref 1).)

In his classic text *The Feynman Lectures on Physics* (ref 2), Richard Feynman describes probability from a scientist's perspective:

> "By the 'probability' of a particular outcome of an observation we mean our estimate for the most likely fraction of a number of repeated observations that will yield that particular outcome."

He goes on to say that such a notion makes sense only if the observations are, as well as we can judge, repeatable. He also clarifies that his use of the word "estimate" conveys that the evaluation is not based on *actual* observations, but rather reflects "what *would* occur in ... *imagined* observations." This clarification separates a probability from a statistical evaluation, and is a key component of probabilistic reasoning. (Throughout this book, we will be asking you to reflect upon what *would* occur in *imagined* ob servations, often then changing some hypothetical conditions then asking what would occur under those new conditions.)

Suppose, for example, that a fair six-sided die is rolled. "Fair" means that each of the outcomes *1, 2, 3, 4, 5,* and *6* is equally likely, so that each has probability $\frac{1}{6}$. *Figure 1* shows a graph of this **probability distribution**. The probability of each outcome is represented by the area of the region corresponding to that outcome.

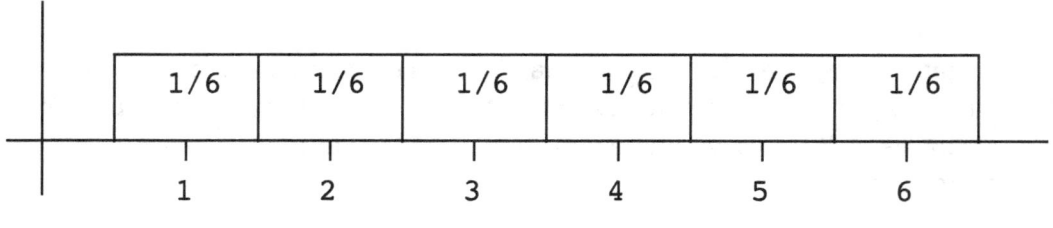

Figure 1

Suppose we actually rolled a six-sided die ten times. One of the authors has just done this, resulting in the outcomes *2, 4, 4, 1, 3, 5, 2, 2, 6, 6*. The result is graphed in various ways in *Figures 2, 3, and 4*.

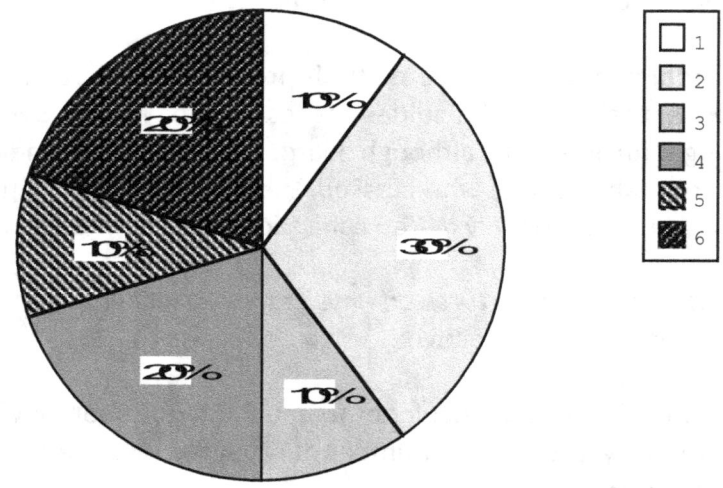

Figure 2

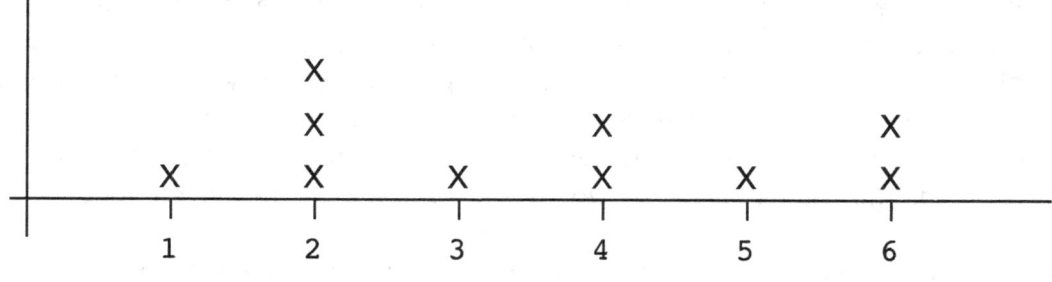

Figure 3

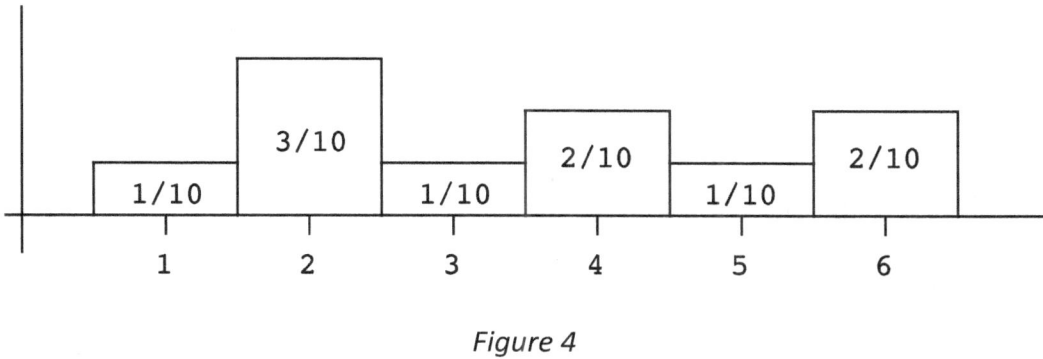

Figure 4

Figure 2 is called a **pie chart**, the size of the pieces depicting the frequencies of occurrence.

Figure 3 is called a **line plot** and provides an easy means of recording data as it occurs.

Figure 4 could be thought of as either a **bar graph** (in which the bars are labeled according to categories and the heights of the bars represent how frequently the various outcomes occurred) or a **histogram** (in which the real number line is partitioned into segments and the areas of the bars represent how frequently the various outcomes occurred). Pie charts, line plots, and bar graphs can be used for categorical data, but histograms can be used only for quantitative data. (*Pay close attention to the distinction between a bar graph and a histogram and then explain why histograms can be used only for quantitative data.*)

The probability distribution of *Figure 1* is an idealization of what the histogram would look like if the die were tossed many times; it is this sort of idealization that Feynman referred to. The graphs in *Figures 2, 3, and 4* are statistical (representing actual data). Actual data is often used to guide one's judgment when evaluating probabilities, and thus, graphs of actual data can be used to help us decide what the graph of a probability distribution might look like, when we wouldn't know otherwise. Alternatively, graphs can be used to compare actual data with a preconceived probability distribution.

It is often useful to *simulate* data through the use of randomly generated numbers. (So, for example, instead of actually rolling a die, you might have a calculator generate random whole numbers for you from 1 to 6.) **Random** means chosen by chance from some particular probability distribution; commonly, people take this distribution to be **uniform** (where every outcome is equally likely). The predicted distribution of rolls of a fair die, depicted in *Figure 1*, is an example of a uniform distribution. In *Homework # 1 (a)*, we will see that it is not easy for people to create random numbers without a device, so we now give you a few methods. If you have access to the internet, you could use a random number generator on a website such as that in (ref 3). Many mathematical or spreadsheet computer programs are equipped with a random number generator. So are many calculators. Tables of random numbers are also easily available. All of these methods rely on some kind of formula to generate the numbers, so the resulting numbers are not truly *random*, but they will work for our purposes. Alternatively, you could draw numbers out of a bag, as we did in *Class Activity 3*; or use the method proposed in *Homework # 2 (c)*.

We would like to stress that when we say something is random, we are referring to a process, rather than the result of that process. For example, suppose we pick 4 digits at random. If our process for picking digits is truly random, than the result of 7777 should occur just as often as 7481 and just as often as 1234. Each possible 4-digit sequence should occur with the same probability. We can't look at a given result, say 1234, and say that the process was not random. A common misconception is that a result of 7481 is more likely to occur in a random process than 1234. Perhaps that misconception occurs because our minds are instead thinking about comparing the likelihood of getting a "mixed up" sequence rather than a special "patterned" sequence of digits. True, it is more likely to get a result that is "something like" 7481 (that is, all mixed up) rather than "something like" 1234 (that is, in a special pattern), but the particular result of 7481 is just as likely in a random process as the particular result of 1234. (Of course, if your house number was 7481 you might think that result is also a special pattern.

Finally, a word about language. When we say "pick a random number", what we actually mean is "randomly pick a number". Really there is no such thing as a "random number", but there are such things as randomly generated numbers. Perhaps we as teachers should avoid saying things like "random number" if we want to counteract the kind of misconceptions in our example above, and instead stress that randomness refers to a process, rather than a result.

Connections to Teaching: Probability Fallacies

Computer simulations may help students avoid or overcome erroneous probabilistic thinking.

NCTM Principles and Standards

The research (Shaughnessy, 1992) identifies several misconceptions that affect the probabilistic judgment of students (and pretty much everyone else too). Here are a few of them.

Availability: Estimating the probability of an event based on how easy it is to call to mind specific instances of that event. This introduces bias based on personal experiences and circumstances. Here are some examples:

- People who have been burglarized will overestimate the crime rate in their neighborhood.

- Because plane crashes get lots of media attention, people think they are more likely to die in an airline disaster than in a car wreck.

- People will estimate that it is more likely for R to be in the first letter of a word than in the third position in a word (even though it *is* more likely third – and so are K, L, N and V – who knew?) because it is easier to call to mind words that begin with R than it is to think up words with R as the third letter.

Conjunction fallacy: Judging that it is more likely that an outcome in A ∩ B occurs than an outcome in just A. Here's an example: Sally is smart and likes to argue. Which is more likely: that Sally is a college graduate, or that Sally is a college graduate who majored in law? A person using the conjunction fallacy would choose that latter based on the descriptors.

Representativeness: Estimating the likelihood of an event based on how well an event represents some aspect of its parent population. So for example, people will say the string TTTTTTTH is less likely than the string TTHTHHHT when a coin is tossed eight times because they expect coin tossing to yield results with about half head and half tails and this event doesn't mirror that aspect of the 'coin string' population. By the way, many students will also think TTTTHHHH is less likely than THTTHTHH because the first doesn't look representative of a random process.

Here are some fallacies that are thought to be consequences of representativeness:

Negative recency effect (or gambler's fallacy): people believe that after a string of losses, they become more likely to win in order to balance things out.

Neglect of sample size: People will ignore the size of the sample when making a judgment about the likeliness of an event. For example, they will say that getting more than 60% heads when tossing a fair coin 10 times is just as likely as getting more than 60% heads when tossing a coin 100 times. Make sure to simulate this (or something like this) in class so you can know the truth.

Base-rate fallacy: People will neglect the rate at which a characteristic appears in the sampled population and instead rely on other things (such as physical descriptions) to judge the likeliness of an event. For example consider this scenario: Suppose a UWO student is selected at random. She is very tall and athletic looking. Is it more likely that she is an elementary education major or on the basketball team? Most people will select the latter as more likely even if nearly 10% of all UWO students are elementary education majors whereas fewer than 0.1% are on the basketball team.

As a teacher, you will be confronted with these faulty ways of reasoning regarding chance – your job is to recognize these misconceptions, draw them out, and address them.

Homework

Experience is the name we give to our mistakes
Oscar Wilde

1) Create a 6-item questionnaire to assess the misconceptions in probability described in the Connections section. Then find some innocent person to answer these questions. Did they show any of the misconceptions? Which ones? Bring the results to class. (Really do this.)

2) Parts (a) and (b) of this exercise should be coordinated with the other students in your class. Parts (a) and (b) hint at why we should not attempt to randomly generate numbers by just thinking them up. Part (c) provides a nice method of randomly generating numbers.

   a) Get everyone in the classroom (or any other large group of people) to pick a random number from 1 to 10. Record the results on an appropriate graph. Is the distribution what you expected? Describe anything unusual you notice.

   b) On one sheet of paper, write down a sequence of 200 H's and T's, attempting to make it look as random as possible. *After* you have done that, on another sheet of paper, actually toss a coin 200 times, and record the sequence of H's (heads) and T's (tails) that result. Compare your two lists. Do you see anything surprising? Look at the lists that some of your classmates created, but without being told which list is which. Can you identify the faked data? Explain how.

   c) If you are not familiar with the binary (base-two) system of representing numbers, read the basics about it. This information is easy to find in an encyclopedia or through a web search. Then explain how you can use 5 coin tosses to generate any random number between 0 and 31. How many tosses would be required to generate a random number between 0 and 1,000?

3) In *Figure 4* we drew a histogram representing the results of ten actual rolls of a six-sided die. Draw three histograms -- the first for two actual rolls of a six-sided die, the second for ten actual rolls, and the third for a hundred actual rolls. How do your histograms compare with the theoretical distribution of *Figure 1*? Compare your histograms with those of several classmates. Describe how the distributions change as the number of rolls increases.

4) There are two jars, one containing 500 white and 500 red marbles, and the other containing 100 white and 1,900 red marbles. (If you have difficulties with the following questions, experiment with smaller collections of marbles to develop your understanding.)

a. Suppose you randomly took two marbles, one from each jar, and switched each to the other jar. Then you randomly took another two marbles, one from each jar, and switched those, and again, and again, many times. What would eventually happen to the number of marbles and proportion of colors in each jar? Explain your reasoning.
b. Suppose instead that you randomly selected one of the 3,000 marbles, and switched it to the other jar, repeating this procedure many times. What would eventually happen to the numbers of marbles and proportion of colors in each jar? Explain your reasoning.

5) Pictured below are 16 identical circles. Close your eyes, and use a pencil to randomly scatter about 100 dots throughout the picture. From this, develop an approximate formula for the area of a circle with a radius of r. (Do not simply state a formula you may have previously memorized, and do not use any measuring devices.) Are the dots that landed outside the bounds of the picture relevant? Do you think the method would have worked as well with only 10 dots? What about with 1 circle instead of 16? (*Note*: The type of method employed in the problem is known as a *Monte Carlo* method, developed by Stan Ulam and John Von Neumann (ref 7).)

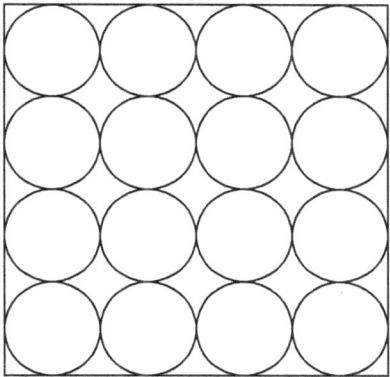

6) The plan is that a northbound bus and a westbound bus will meet in Grover's Corners every hour at the hour, but of course neither one ever shows up exactly on time. In fact, they arrive randomly between 10 minutes before the hour and ten minutes after the hour, with a uniform distribution. The rule they use is that if a driver arrives and does not see the other bus, he or she is to wait 5 minutes, then proceed. Suppose you are riding the northbound bus toward Grover's Corners, scheduled to arrive at 1:00 pm. What is the probability that you will get there in time to catch the 1:00 pm westbound bus (and not have to wait for the 2:00 pm bus)?

7) The following problem is motivated by an old observation that certain portions of books containing log tables (a useful arithmetic tool before calculators were invented) were more worn from use than others. If you are interested in reading more about this, see a discussion of Benford's Law in (ref 5). A more mathematically sophisticated presentation appears in (ref 6).

a) Draw a histogram representing how you *think* the integers from 1 to 9 would be distributed, if you were to just walk around observing numbers as you went about your everyday life.

b) Put your conjecture from a) to the test. As you go about your everyday life, keep a record of the first nonzero digit of each number you see. For example, if you saw the date October 6, 2005 written "10 / 06 / 2005" you would record the numbers 1 (from 10), 6 (from 06) and 2 from (2005); so you will just be writing down numbers between 1 and 9. Collect these numbers from anywhere you can find them -- newspapers, books, street signs, etc.... Try to collect about 1,000 numbers (or work together with your classmates to achieve this goal). Plot this data and compare it with your graph from part a). Was your prediction accurate? Can you explain what is going on?

Class Activity 5a: The Best Answer

The temptation to form premature theories upon insufficient data is the bane of our profession.

Sherlock Holmes (Sir Conan Doyle)

Materials needed: a large bag (about 2 lbs.) of peanut M&M's;
a glass jar or clear plastic container.

Empty the M&M's into the jar. Then everyone should, individually, carefully estimate the number of M&M's in the jar. After all estimates have been made, they should be written on the chalkboard. Then the groups within the class should compete with each other, each group using this list of estimates to make its best guess as to the true value. When this is done, the groups should take turns presenting and explaining their solutions. Afterwards, the various methods can be discussed and debated. Ultimately, the M&M's should be counted. Maybe the group whose answer was closest will get to eat them?

Class Activity 5b: Hanging in the Balance

Your task is to put weights on the numbered pegs to keep the system balanced.

1. Find several different arrangements of 3 weights so that the system will balance. Make a general conjecture about all the arrangements that will balance the system.

2. If you had one weight 1 to the right and two weights 2 to the left, predict where you would need to place a fourth weight to balance the system.

3. Find several ways to balance four weights where only one weight is on the left side. Make a general conjecture about all the arrangements of this type that will balance the system.

4. Predict some arrangements that will balance for 5 weights. Make a general conjecture about all the arrangements of 5 weights that will balance the system.

5. Suppose there are 4 cats with mean weight of 10 pounds. You know the weights of 3 of the cats. Those weights are: 4 pounds, 7 pounds, and 15 pounds . How can you use the balance to figure out the weight of the 4th cat? (Don't do a calculation. Figure it out with the balance!).

6. Use the balance to figure out this problem: You have taken four math quizzes. Your scores were 82, 67, 85, and 72. What do you need to score on the 5th quiz for your average quiz score to be 75? (Don't do a calculation. Figure it out with the balance!).

Read and Study: Measures of Center

I am only an average man but, by George, I work harder at it than the average man.
Theodore Roosevelt

The November 7, 2000 U.S. presidential election remained undecided for more than a month after the last poll had closed. The outcome hinged on the state of Florida, where the vote count was too close to call. A variety of circumstances had created lots of problems: votes were mistakenly cast for the wrong candidate, ballots were cast but for which the counting machines registered no vote (undervotes), and ballots were cast for which the machines registered votes for more than one candidate (overvotes). On December 9, amidst a legal turmoil, the U.S. Supreme Court stopped the hand recount that was underway, deciding the election in favor of George W. Bush, who at the time held a 537 vote lead over Al Gore.

Subsequent privately-sponsored audits of Florida's votes remained inconclusive; the winner depended on the standards used in counting the votes (ref 1, 2). Furthermore, the auditors' handling of the punchcards sometimes resulted in new chads being torn, thus changing the vote that was being counted. The search was kind of like the 1966 film *Blowup*, in which a photographer, attempting to discover whether he has photographed a murder, blows the picture up larger and larger in search of a gun, only to find it so grainy as to be indecipherable.

Any measurement is subject to uncontrollable sources of error. Recognizing that even his own meticulous measurements fluctuated from one reading to the next, astronomer Tycho Brahe was, in the late 1500's, perhaps the first scientist to combine *repeated* measurements of the same quantity in order to produce a single reliable value (ref 3). (You may recall from *Section 1* that Tycho's student Johannes Kepler used his data to develop a mathematical description of the planetary orbits.)

Since that time, scientists have used a variety of methods to boil down a collection of data into a single representative value. As it was with Tycho, sometimes the goal is to identify a "best" value from a set of repeated measurements of the same thing. Other times the goal is to use a single value to represent some aspect of a population of many different things. (For example, the *World Book Encyclopedia* (ref 4) states, "The average weight of a baby born at term is 7 1/2 pounds." *What does this mean? Are they referring to a particular baby*?)

Two representative values have emerged to the forefront of modern statistics: the **mean**, defined as the *average* of all the values; and the **median**, defined essentially as the *middle* value upon listing the data in order from smallest to largest. The median can be chosen as any value that is at least as big as at least half of the data, and at least as small as at least half of the data (ref 5). *Can you think of an example of a data set for which there is more than one median? Explain.* Another value used occasionally is the **mode**, defined as the most frequently appearing among the data values.

Which of these is "better" to use depends on which type of questions you are considering. The mode, for instance, can be used for categorical data, but the mean and the median do not make sense in that context. *Why? If you wanted to know how much money you would make working at Microsoft Corporation, would it be more honest to look at the mean income of its employees, or the median income? Explain.*

The mathematics surrounding the mean is the most fully developed. It allows us to describe data sets in a similar way to how a physicist describes physical objects. (*Class Activity 5b* was written to help you see this connection.) It also leads to a convenient way of describing errors in measurement and variation among the members of a population, the topic of *Section 7*. Furthermore, analysis of means has become a standard method of using samples to make inferences about larger populations. This will be the focus of the last several sections of the book.

Above, when we defined the mean, it was for a set of data. If that data merely represents a *sample*, then it is called the **sample mean.** If our variable is represented by x then the usual notation for a sample mean is $\bar{x}$ (pronounced *x*-bar). However, if the mean we are talking about represents the entire *population*, then it is called the **population mean**. The usual notation for a population mean is μ. This is the lower-case Greek letter mu (pronounced "mew"), which is the Greek equivalent of the letter *m* for mean.

A **census** is the collection of information from every member of a population. The origin of the word *census* is the Latin *censere*, "to tax." A census was used in ancient Rome to determine the taxable population as well as the number of potential soldiers (ref 6). Every 10 years, the U.S. Census Bureau conducts (well, *attempts* to conduct) a census of the people living in the United States. Among the information gathered are people's ages. The average of everyone's ages would be a *population mean*. The Bureau collects additional information from a sample of about 1/6 of the population. The average of these respondents' ages would be a *sample mean*. (For more information about the U.S. Census Bureau and its methodology, see ref 7.)

Recognizing that it does not successfully obtain information from everyone, the Census Bureau has proposed integrating its direct counts with its large-sample surveys for the official results. Sampling methods have indicated that minority ethnic populations tend to be undercounted more than whites, and thus are underrepresented. This has led to a battle between Democrats, who generally favor the inclusions of sampling results, and Republicans, who generally oppose their inclusion. The Supreme Court ruled in 1999 that although sampling methods can be used for some purposes, they cannot be used to apportion Congressional Representatives among the 50 states (ref 8).

Connections to Teaching: Relating Mean and Median

In grades 6-8 all students should find, use and interpret measures of center and spread including mean and interquartile range.
NCTM Principles and Standards

One *idea* of mean that can be taught is that the mean is like "evening out" the data. That is, it is what each data point would get if you evened them out. It can also be thought of as the balance point of a data distribution.

NCTM's online *Illuminations* materials has an applet that can be used for developing an understanding of how the mean and median are related to a data set. Access the applet by visiting: https://illuminations.nctm.org/Activity.aspx?id=3576 Read the applet's instructions and play around with the applet (under Activity) to get a feel for what the applet does and what information it provides. Then use it to do the *Exploration* found under the appropriate tab on the webpage. *What are some things that students could learn by doing the exploration provided using the applet? How do these ideas help students understand how the mean and median relate to a distribution of data?*

Homework

Many of life's failures are people who never realized how close they were to success when they gave up.
Thomas A. Edison

1) Find a statistics applet that will allow you to enter data, and will calculate the mean and the median of that data for you. Use it to create the following two data sets, each comprised of exactly ten integers that are between 0 and 100:
   a) Data Set A: A set with a mean of 50 and a median of 40.
   b) Data Set B: A set with the largest possible difference between the mean and the median.

2) Consider the following graph of the number of people in 6 children's families.
   a) Figure out how you can *use the graph* to find the mean family size, *without doing any calculations*. Explain how you did it.
   b) Explain why this graph should not be called a "bar graph."

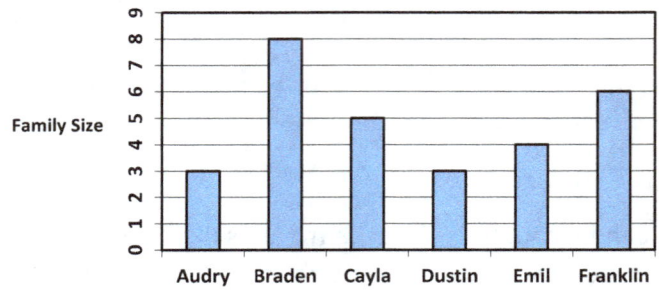

3) Below is a line graph showing the number of pets in 16 households. Figure out how you can use the graph to find the mean number of pets per household, without doing any calculations. Explain how you did it.

```
            X   X
    X       X   X
    X       X   X
    X       X   X       X
    X       X   X       X   X   X           X
    _____
    0   1   2   3   4   5   6   7   8
            Number of Pets
```

4) Last semester I had two sections of Math 217 that took an exam. Section A had 25 students, and the mean score was 65 points with a median of 80 points. Section B had only 15 students and a mean of 75 points and a median of 70 points. If possible, determine the mean and median scores over all 40 students.

5) If possible, find data sets with the following properties:
a. $n = 10$, $\bar{x} = 20$, median = 30, min = 0
b. $n = 6$, $\bar{x} = 12$, min = 4, max = 16

6) A data set consists of 10 whole numbers. The median is 4 and the mean is 5. What are the largest and smallest possible values for the maximum?

7) Suppose you were to toss a coin a few times. Considering each toss as a separate experiment, find a way of assigning numerical values to replace *head* and *tail*, so that the mean of these values would equal the proportion of *heads* among all the tosses.

8) An article in a 2004 issue of the University of Wisconsin Oshkosh student newspaper, the *Advance-Titan*, implied that Wisconsin governor Jim Doyle might have been influenced by campaign donations when appointing members to the UW-system Board of Regents (ref 12). The basis for the allegation was that the 8 members appointed by Gov. Doyle had

contributed an average of $4,785.38 while the 9 members who were *not* appointed by Doyle (they were still serving a previous appointment) had contributed an average of only $935. The breakdown of donations is reported as follows: among the 8 members appointed by Doyle, one contributed $25,750, one contributed $9,100, one contributed $3,333, one contributed $100, and four did not contribute; among the 9 members *not* appointed by Doyle, one contributed $8,000, one contributed $250, one contributed $220, and six did not contribute. (By the way, have you spotted the error in one of the reported calculations?) What is it about this data that makes the mean (average) an unreasonable basis for comparison? Would the median be affected in the same way? Check it out by comparing the median contributions.

9) [Note: This problem requires you to collect data from your classmates.] In a previous homework problem you drew a histograms representing the distribution of the individual outcomes of 2 actual rolls, 10 actual rolls, and 100 actual rolls of a die. Use these histograms to calculate the means of each distribution. Are these sample means or population means? Next, also record the means obtained by each of your classmates. What major observation can you make regarding these values? Does the number of rolls seem to have an effect on the means?

Class Activity 6a: The St. Petersburg Lottery

A fair coin is flipped until the first tails appears. You win $1 if the tails appears on the first toss, $2 if the tails first appears on the second toss, $4 if the tails first appears on the third toss, $8 if the tails first appears on the fourth toss, etc. If you were to play this game, on average how much can you expect to win? How much would you be willing to pay to be able to play this game?

Class Activity 6b: The Problem of Points

Players A and B are playing a sequence of fair games against each other. So far, A has won 3 games and B has won 4. A $100 prize will go to the first one to win a total of 5 games. If the play is interrupted right now, how should the $100 be divided up?

Read and Study: Expected Value

Often, as in the case of a scientist's repeatable experiment, samples do not come from a tangible population, but rather are generated according to some probability distribution. The mean of how we might *expect* our data to turn out, based on the underlying distribution, is called the **theoretical mean**. It is an average of the values that could occur, weighted by their likelihoods according to the distribution.

Suppose, for example, that you wanted to know the average outcome of rolling a fair die. The mean of this distribution is 1/6 (1) + 1/6 (2) + 1/6 (3) + 1/6 (4) + 1/6 (5) + 1/6 (6), which equals 3 ½. So 3½ is the *theoretical mean*. If you actually rolled a die 100 times, and took the average of the outcomes, that value would be a *sample mean*.

The theoretical mean is sometimes called the **expected value**, since it represents how we might *expect* our data to turn out, based on the underlying distribution. In the context of playing games, if we have a theoretical probability distribution for the payouts associated with the various outcomes of the game, the mean of this distribution, or the **expected value** of the game, is how much you expect to win (or lose) on average, if you played the game a large number of times.

Suppose, for example, you play a simple game where you flip a coin, with the following payoffs: if you call the coin correctly, you win $1; if you don't call it correctly, you lose $1. Then the expected value of this game is zero. That is, in the long run, we expect to win as much money as we lose. Games with an expected value of zero are called **fair games**.

Gambling games such as the lottery or casino games are never fair; the expected value of these games is always negative… that's how the people running the game make their money. Typically in casino games, the expected value is only slightly negative, so that people win quite often, but it the long run, they will lose more than they win. However, in lottery games, the expected value is often very negative. Here the allure for playing is that the cost to play is relatively small compared to the enormous payoff.

In the first class activity, you worked on what is known as the St. Petersburg Paradox, even though it has very little to do with St. Petersburg, Russia. This problem was posed by Swiss mathematician Nicholas Bernoulli in1713 and solved by his cousin Daniel Bernoulli in 1738. Daniel worked for a time with fellow Swiss mathematician Leonard Euler at the university in St. Petersburg, and Daniel published his solution in the *Commentaries of the Imperial Academy of Science of Saint Petersburg.* What makes this problem a paradox is that even though the expected value is infinite, a rational person would not be willing to pay very much money to play this game.

In the second class activity, you worked on a specific case of what is called the Problem of Points, *Two people are playing a sequence of fair games against each other; a prize will go to the first one to win a certain number of games. If the play is interrupted before it is over, how should the prize be divided up?*

The Problem of Points dates back to the 1600s. The systematic study of the theory of probability is thought to have originated with the solution of this problem by Pascal and Fermat. In 1654 they wrote letters back and forth in which they solved not only this problem but laid the groundwork for probability as a mathematical discipline.

Homework:

1. The Wisconsin Lottery offers a game called "Pick Three", where you have to correctly pick a three-digit number in sequence. The game costs $1 to play, and if you win, the payout is $500. What is the expected value of this game?

2. Suppose you pay $1 for the right to play this game: you roll a die and you win if you roll a 4, otherwise you lose. What should be the payoff for winning this game for it to be a fair game?

3. A Roulette wheel contains the numbers 1, through 36, 0 and 00. Thus it has 38 places where the ball can stop. You choose a number or set of numbers, the wheel spins, and if the ball stops on your number, you win. If choose a single number, and the ball stops on your number, you win $36 for every dollar you bet. If you pick odd (or even), and the ball stops on an odd (or even) number from 1 to 36, you get double your money back. (In Roulette, neither 0 nor 00 are considered even). If you pick one-third of the numbers (numbers 1-12, 13-24, or 25-36), and one of your numbers comes up, you win triple your money back.
    a. Ursula has $3 to bet on Roulette. She decides to bet $3 on 00. What is the expected value with this strategy?
    b. Tim also has 3 dollars to bet on Roulette. He decides to bet $1 on the first third, $1 on the second third, and $1 on the last third, all on the same spin on the Roulette wheel. What is the expected value with this strategy?
    c. Sue also has 3 dollars to bet on Roulette. She decides to bet $1.5 on evens and $1.50 on odds, all on the same spin. What is the expected value with this strategy?
    d. Raffy also has 3 dollars to bet on Roulette. He decides to bet $1 on evens for three consecutive games. What is the expected value for this strategy?

4. Suppose your probability of winning a game is p. If you win, you get A dollars. If you lose, you must pay B dollars. What is the expected value of this game?

5. Powerball is a combined large jackpot game and a cash game. They draw five white balls out of a drum with 59 numbered white balls and one red ball out of a drum with 35 numbered red balls. It costs $2 to play, and you try to pick the 5 white numbers (in any order) and the one red number. The prize you can win depends on how many of the numbers you pick

correctly. According to the Powerball website, the odds of winning each prize is provided below.

Match	Prize	Odds
●●●●● + ●	Jackpot	1 in 175,223,510.00
●●●●●	$1,000,000	1 in 5,153,632.65
●●●● + ●	$10,000	1 in 648,975.96
●●●●	$100	1 in 19,087.53
●●● + ●	$100	1 in 12,244.83
●●●	$7	1 in 360.14
●● + ●	$7	1 in 706.43
● + ●	$4	1 in 110.81
●	$4	1 in 55.41

a. Assuming an average jackpot of $150 million, what is the expected value of playing Powerball?

b. How large does the Jackpot need to be in order for Powerball to be a fair game?

6. We've invented a new carnival game. At the beginning of the game, a gumball machine is stocked with 8 gumballs. Four are red and four are blue. To play, the player pays a dollar, turns the crank, and gets out one gumball. If the gumball is red, the player wins the game and gets a cash payout. If the gumball is blue, the player gets to eat it, and can pay another dollar to try again, repeating until they get a red gumball. How much should the payoff be for getting a red ball for this to be a fair game? (Neglect the cost of the gumballs).

7. Here's another game we've made up. You roll a pair of dice and taking their sum. If you roll an odd sum, the game is over. If you roll doubles of any number, you win $10 and the game is over. If you roll an even sum (other than doubles) you may continue rolling. A casino needs to set a price for playing this game so that they make a profit in the long run. How much should the game cost to play?

8. *The Passion of the Christ*, a film which explicitly portrayed Jesus' final twelve hours, was released in February of 2004. Within a month, two people had died while watching the movie. Do you find this unusual? Estimate the number of deaths you would expect to have happened. You can look up some relevant facts about the movie online at these sites:
http://www.imdb.com
https://www.boxofficemojo.com/release/rl3781789185/

Class Activity 7a: Shufflehall

We've created a new game: You push a chalkboard eraser down the hall to the goal line, which is 24 feet away. Each player gets 5 tries. There are two possible winners of a shufflehall game: the most *accurate* player, and the most *consistent* player.

Who is the most accurate shufflehall player in our class?

Who is the most consistent shufflehall player in our class?

Class Activity 7b: How Many Tanks?

You may have to fight a battle more than once to win it.
Margaret Thatcher

Your group is going to assemble a fleet of tanks, because you are at war with the one of the other groups in the classroom. Of course you give each tank a serial number for identification, but as you will discover, this gives the enemy an ability to estimate the size of your fleet…

What to do: Each group should (without letting the other groups overhear) select a minimum and a maximum integer, representing the smallest and largest serial numbers on the tanks in your fleet. (For example, if you selected 200 as your minimum and 299 as your maximum, your fleet would have 100 tanks.) Then use a random number generator to identify ten tanks that were "captured" from your fleet, give those ten numbers to another group, and get their ten numbers. (You're pretending they captured ten tanks from your fleet, and you captured ten tanks from their fleet.) Using the ten serial numbers you were given, try to guess how many tanks the other group had in its fleet. Prepare to discuss your methods of analysis.

Note: This type of analysis was actually useful to the United States in estimating numbers and production rates of German tanks, trucks, bombs, etc… during World War II. Some historical details are provided in (ref 1). The advantages and disadvantages of various methods are discussed in (ref 2), as "the locomotive problem," and especially in (ref 3), where the discussion is tailored for a school classroom.

Read and Study: Measures of Spread

Saying what we think gives a wider range of conversation than saying what we know.
Cullen Hightower

There are many ways to describe the *spread*, or variation, within a set of values. Probably the simplest is the **range**, the difference between the maximum and minimum values. The median can be used to break the range into two halves. This idea can be extended: the range could be broken into 100 equal pieces, dividing the data into *percentiles*, or the range could be broken into 4 equal pieces, dividing the data into *quartiles*, and so on.

Suppose you separated a set of data into its 4 quartiles, and labeled the 3 points separating the quartiles as q_1, q_2, and q_3, in increasing order. The median is the same as q_2. These 3 values, together with the minimum and maximum values, are called the *5-number summary*. We use these points to define something called the **interquartile range (IQR)**, $q_3 - q_1$. Why might it make sense to report the interquartile range of a sample, rather than the entire range?

The 5-number summary can be graphed in a simple way using a **box plot**, also known as a **box-and-whiskers plot**. An example is shown below.

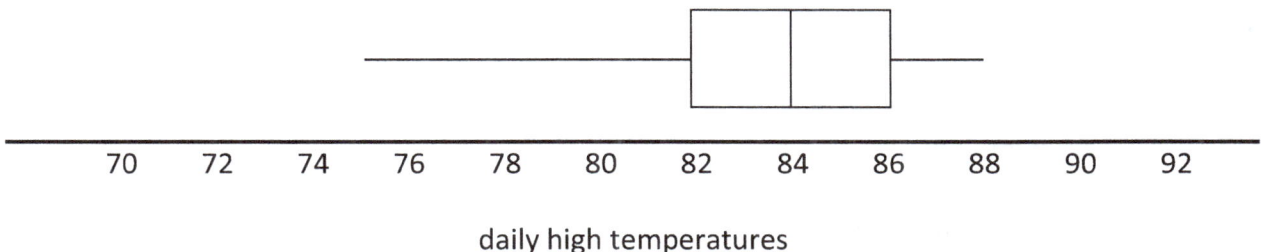

daily high temperatures

This plot represents the daily high temperatures (in degrees Fahrenheit) in Oshkosh, Wisconsin for the 31 days of July, 2010 (ref 4). From the plot, we can see that the *minimum* daily high was 75º, q_1 was 82º, q_2 (the *median* temperature) was 84º, q_3 was 86º, and the *maximum* daily high was 88º. The "whiskers" extend to the minimum on the left and the maximum on the right, so the *range* is the length of the whole thing. The *interquartile range* is the length of the 'box' part. The *median* (q_2) is designated by the bar in the middle of the box. *Based on the data above, approximately how many days had a high temperature above 82º? Explain.*

The purpose of the box plot is to allow you to quickly get an idea how the data is spread out; therefore, it is important to accurately space the numbers on the scale below the plot, as they would appear on a number line.

We commented earlier that the 16th-century astronomer Tycho Brahe was aware of the fluctuation inherent in his measurements. However, there is no evidence of any study of such errors until about 30 years after Tycho's death, when Galileo postulated that errors in a measurement would be symmetrically distributed about the true value, with small errors being more likely than large ones. Without explicitly identifying the best representation of the true value, he did indicate that it should be the one to minimize the observational errors (ref 5, 6).

In one sense, observational errors are minimized if the *median* is chosen as the representative of the true value: the median happens to be the value minimizing the sum of the distances of the observations from said value. In another sense, observational errors are minimized if the *mean* is chosen as the representative: the mean happens to be the value minimizing the sum *of the squares* of the distances of the observations from said value (ref 7). Because of both its mathematical convenience (recall from *Section 4* that for this reason we tend to favor the mean) and its connections to physics, it is the squares of the distances from the mean, as opposed to the actual distances from the median that we more commonly use as a measure of spread.

The *standard deviation* of a sample is defined as follows: take the "average" of the squared distances from the individual observations to the sample mean, and then take the square root of that – except that, when taking the average, rather than dividing by the number of observations, divide by 1 less than this number. (We won't worry in this book about why statisticians subtract 1 from the number of observations, other than to say it is essentially to make things come out the way you would expect them to.) Let's present this definition more carefully. Where

- n is number of observations (the sample size)
- x_i denotes the ith observation
- and $\bar{x}$ is the sample mean

we define the **standard deviation** of the sample to be

$$\sqrt{\frac{(x_1 - \bar{x})^2 + (x_2 - \bar{x})^2 + \cdots + (x_n - \bar{x})^2}{n-1}}.$$

Using *Class Activity 6B* as an example, suppose the other group drew the tank sample 5, 15, 36, 64, and 100 from your bag. The sample mean would be 44 and the *standard deviation* of this sample would be

$$\sqrt{\frac{(5-44)^2+(15-44)^2+(36-44)^2+(64-44)^2+(100-44)^2}{5-1}}.$$

The main purpose of taking the square root is so that the unit of measurement will be the same as that for the data itself. For instance, if your data was measured in feet, then, since each of the deviations got squared, taking the square root allows the standard deviation to be measured in the original units, feet.

The Common Core State Standards does not introduce *standard deviation* until the high school level. Prior to that (beginning in grade 6) the standards advocate both the *interquartile range* and the *mean absolute deviation* as measures of variation. They define the **mean absolute deviation** (we'll abbreviate this as **MAD**) as "a measure of variation in a set of numerical data, computed by adding the distances between each data value and the mean, then dividing by the number of data values." It is similar in spirit to the standard deviation, but simpler to calculate. For example, using the above numbers for which we just calculated the standard deviation, let's now calculate the MAD:

$$\frac{|5-44|+|15-44|+|36-44|+|64-44|+|100-44|}{5}$$

To summarize the last several ideas, the MAD is a measure of how far away the data is, on average, from the mean, and the standard deviation is a more sophisticated but similar measure. Along with the range and the interquartile range, they measure the spread in a data set. In a later section we will say more about the fundamental role played by the standard deviation in modern statistics.

Connections to Teaching: Determining Outliers

[Middle Grades students should see that] the mean is very sensitive to the addition or deletion of one or two extreme data points, whereas the median is far less sensitive to such changes.
                                        NCTM Principles and Standards

Middle grades students study MAD, range, and interquartile range (IQR) as measures of spread. (As mentioned in the *Read and Study*, standard deviation is typically a high school topic.) As a middle grades teacher, you should focus on the idea of an outlier in a set of data, and help your students to understand that when outliers are present, the IQR is the better measure of the data's spread. (The median is the better measure of center in that case too.) An **outlier** is an individual data point that lies well outside the overall pattern of the data.

If the data is clustered together, then it doesn't take much to be an outlier.

```
              X
              XX
              XXXXX
              XXXXXX        X
        ─────────────────────────
        0 1 2 3 4 5 6 7 8 9 10
```

If the data is spread out, then for a data point to be an outlier it must be *really* far away from the bulk of the data.

```
           X                      X
X          XXXX         X  X          X           X              X
0 1 2 3 4 5 6 7 8 9 10 11 12 13 14 15 16 17 18 19 20 21 22 23 24 25 26 27 28 29 30 31
```

One way statisticians define what we mean by "well outside" the pattern of the data, is to use the **1.5 × (IQR) criterion**. This criterion says that a data point is an outlier if it is either

> more than 1.5 × (IQR) above q_3
> or
> more than 1.5 × (IQR) below q_1.

Use the 1.5 × (IQR) criterion to see if either of the data sets above has an outlier.

Are there any outliers in the boxplot showing the daily high temperatures? Explain how you can use the box to think about this.

Homework

Once you learn to quit, it becomes a habit.
Vince Lombardi

1) Do all the italicized things in the *Read and Study* and *Connections* sections.

2) Consider the data set shown at the top of this page. Calculate the Mean, Median and MAD and IQR for this distribution. If the highest data point (at 10) was determined to be an outlier and omitted from the data set, how would this change the Mean Median, MAD and IQR? If the highest data point (at 10) was actually supposed to be at 50, and was kept in the data set, how would this change the Mean, Median, MAD and IQR? Which measures of center and spread are the most resistant to outliers?

3) For this problem you will need a pair of 10-pound dumbbells (or something similar, like two heavy backpacks)
   a) Pick up the dumbbells, one in each hand, hold them out to your sides at arm's length, and spin slowly around in a circle. While continuing to spin, pull your arms in to your chest. What happens and why? Can you figure out a connection to the concept of standard deviation? (Isn't this fun?)
   b) Next, get rid of one of the dumbbells, and hold the other one out to one side at arm's length and spin around again. What difference do you notice in your spin compared to when both arms held weights outward, and what causes this difference?

4) One of the authors has just rolled a fair die 100 times and a loaded die 100 times. The table below represents the results for those two dice. We also created a third data set by relabeling the data values on the loaded die.

Value on die	Fair die Frequency	Loaded die Frequency	Relabled loaded die Frequency
1	19	1	1
2	22	6	5
3	14	3	84
4	17	5	6
5	13	1	3
6	15	84	1

   Draw three histograms corresponding to these three data sets. Then, by studying the histograms (and *not* by performing any calculations), decide which of these data samples has the smallest standard deviation and which has the largest. Explain your reasoning.

5) Toss a nickel about 100 times, on each toss recording a "0" if it lands tails and a "1" if it lands heads.
   a) Calculate the standard deviation and the MAD for the data you have collected for your nickel.
   b) Now do the same for a quarter, but instead record a "-1" if it landed tails, and a "1" if it landed heads. You should end up with a larger standard deviation and a larger MAD than you did for the nickel. Does this imply that quarters are more erratically behaved than nickels?

6) Marisa counted the number of birds that came to her feeder during breakfast for 6 consecutive days. On the first four days, she saw 10, 8, 10, and 14 birds. The data for the last two days she lost. But over the entire 6 days, she remembers that the mean number of birds she saw was 10 birds. If the mean absolute deviation over the 6 days was 3 birds, how

many birds could she have seen on each of the other 2 days? What is the largest possible mean absolute deviation that you can make by choosing the last two numbers of birds, while keeping the mean over the six days at 10 birds?

7) The two standard college entrance exams in the U.S. are the SAT and the ACT. Scores on the ACT can range from 1 to 36. Scores on the SAT can range from 600 to 2400. At the end of this homework set we have provided you with a list of all ACT and SAT scores from 2011. Notice that the frequencies are reported in slightly different ways.

   a) Draw the box plots corresponding to each of these two data sets.

   b) The standard deviation among SAT scores is much larger than the standard deviation among ACT scores. Likewise for the MAD. Why does this make sense? (You don't need to do any calculations.

   c) Sammi took both tests that year. Her ACT score was 28, and her SAT score was 1800. Which was the better score?

8) People or groups will need to work in pairs for this problem. Each pair will need 3 large sheets of paper, a pencil, and a few sheets of newspaper for padding. Begin by placing a few layers of newspaper on the floor, for padding. Then place two of your large sheets of paper, stacked together, on top of the newspaper. *Person A should now close his/her eyes for the next couple of minutes.* Meanwhile, Person B should draw an **X** in a random location on the top sheet of paper (but not too close to the edge, and being careful not to leave an imprint on the second sheet); then, standing and holding the pencil at shoulder height, he/she should attempt to drop the pencil on top of the **X**, ten times. (This should result in 10 holes through both sheets of paper.) Upon completing this procedure, remove and hide the top sheet of paper, so that the pencil holes in the bottom sheet are visible but the **X** has been removed. Person A may now open his/her eyes. (Yes, John made up this problem too.)

   a) Upon studying the pencil holes, Person A should try to guess the location of the **X**. Explain your method. (Note: Law enforcement agencies are sometimes confronted with situations such as this, trying to identify the location of a bomb that exploded by analyzing the destruction, or the location of a serial killer's residence based on where the killings occurred; for a profile of a criminologist who has developed some of these techniques, see ref 10.)

   b) What is different about the nature of this data compared with any data for which we have previously defined a "mean"? Can you think of a natural way to extend our notion of a mean for this type of data?

   c) On your third sheet of paper, draw another **X** and repeat the process of attempting to drop your pencil on top of it, but this time, aiming from about a foot above the paper rather than from shoulder height. (Nobody needs to close his/her eyes this time.) What is different about this pattern of holes compared with the previous one, and what accounts for the difference?

9) Earlier in the book, while working with scatterplots, we discussed the *correlation coefficient*, a measure of the strength of a linear relationship between two variables. By now we have developed enough machinery to give you the formula:

$$r = \frac{1}{n-1}\sum_{i=1}^{n}\left(\frac{x_i - \bar{x}}{s_x}\right)\left(\frac{y_i - \bar{y}}{s_y}\right).$$

You might need your instructor's help reading this formula before you try to answer the questions below. We'll start you out by defining the pieces:

- n is the number of observed sample pairs;
- (x_i, y_i) is the ith data value of the pair of variables (X, Y);
- $\bar{x}$ and $\bar{y}$ are the sample means for the variables X and Y;
- s_x and s_y are the standard deviations for the variables X and Y
- the Greek letter capital sigma is a summation symbol (telling you to add repeatedly as the index i changes from 1 to n).

Don't get worried – we won't be using a lot of symbols this complicated in the text. But sometimes we can really understand a concept only by studying the technical details. Our purpose in this problem is to give you some practice thinking in the technical language of mathematics.

a) When drawing a scatterplot, you get to choose which variable is X and which variable is Y. Does your choice affect the value of r? Why or why not?

b) Does r have any units associated with it? (For example, if the variable X is measured in feet and Y is measured in seconds, what is r measured in?)

c) Explain why $r = 1$ if all of the sample data points lie exactly on a line with positive slope. Likewise, explain why $r = -1$ if all the data points lie exactly on a line with negative slope.

10) Read the Common Core State Standards for Statistics and Probability in Grade 6. Identify exactly which parts of these standards have been addressed so far in this book. For each topic of competency mentioned, specify an example or activity (from the Class Activity, Read and Study, Connections, or Homework) that addresses it. Also identify any parts of these standards have not (yet) been addressed in this book.

ACT composite scores, frequencies, and percentiles

ACT	freq	% <=	ACT	freq	% <=
1	0	1	19	109361	41
2	0	1	20	112403	48
3	1	1	21	112763	55
4	4	1	22	108853	62
5	13	1	23	101879	68
6	39	1	24	94412	74
7	89	1	25	83561	79
8	326	1	26	71634	83
9	961	1	27	62086	87
10	3490	1	28	53129	91
11	13195	1	29	42161	93
12	33154	3	30	35834	95
13	54312	7	31	27251	97
14	70462	11	32	20319	98
15	82415	16	33	14283	99
16	92738	22	34	9154	100
17	100787	28	35	4204	100
18	107135	34	36	704	100

SAT composite scores, frequencies, and percentiles

SAT	N	% <	SAT	N	% <	SAT	N	% <
600	180	0	780	1278	1	950	4430	3
610	77	0.5	790	1322	1	960	4777	3
620	109	0.5	800	1485	1	970	5134	4
630	122	0.5	810	1620	1	980	5419	4
640	195	0.5	820	1906	1	990	5563	4
650	187	0.5	830	1943	1	1000	6025	5
660	273	0.5	840	2044	1	1010	6392	5
670	267	0.5	850	2259	1	1020	6702	5
680	376	0.5	860	2475	1	1030	7210	6
690	446	0.5	870	2631	2	1040	7499	6
700	459	0.5	880	2826	2	1050	8004	7
710	649	0.5	890	3082	2	1060	8602	7
720	696	0.5	900	3085	2	1070	8874	8
730	670	0.5	910	3384	2	1080	9331	8
740	831	0.5	920	3601	2	1090	9834	9
750	855	0.5	930	3939	3	1100	10295	9
760	1047	0.5	930	3939	3	1120	11231	11
770	1139	0.5	940	4175	3	1130	11672	11

SAT	N	% <	SAT	N	% <	SAT	N	% <
1140	11994	12	1570	18798	59	2000	6709	93
1150	12588	13	1580	18569	61	2010	6381	93
1160	13087	14	1590	18033	62	2020	5957	94
1170	13371	14	1600	18133	63	2030	5849	94
1180	13948	15	1610	17939	64	2040	5683	95
1190	14233	16	1620	17523	65	2050	5315	95
1200	14752	17	1630	17516	66	2060	5041	95
1210	15296	18	1640	17033	67	2070	4926	96

1220	15782	19	1650	16745	68	2080	4536	96	
1230	16161	20	1660	16650	69	2090	4403	96	
1240	16600	21	1670	16142	70	2100	4340	96	
1250	16945	22	1680	15794	71	2110	4028	97	
1260	17252	23	1690	15578	72	2120	3890	97	
1270	17396	24	1700	15451	73	2130	3684	97	
1280	17688	25	1710	15108	74	2140	3370	97	
1290	17989	26	1720	15055	75	2150	3259	98	
1300	18648	27	1730	14637	76	2160	2970	98	
1310	18815	28	1740	14313	77	2170	2837	98	
1320	19002	29	1750	14015	78	2180	2739	98	
1330	19352	30	1760	13674	78	2190	2555	98	
1340	19425	32	1770	13211	79	2200	2475	98	
1350	19240	33	1780	13035	80	2210	2274	99	
1360	19901	34	1790	12856	81	2220	2183	99	
1370	20053	35	1800	12379	82	2230	2028	99	
1380	19760	36	1810	12125	82	2240	1948	99	
1390	20082	38	1820	11758	83	2250	1778	99	
1400	20193	39	1830	11583	84	2260	1545	99	
1410	20067	40	1840	11136	84	2270	1555	99	
1420	20170	41	1850	10752	85	2280	1359	99	
1430	20389	42	1860	10524	86	2290	1221	99	
1440	20139	44	1870	10207	86	2300	1235	99.5	
1450	20418	45	1880	9920	87	2310	948	99.5	
1460	19974	46	1890	9608	88	2320	1055	99.5	
1470	20251	47	1900	9293	88	2330	767	99.5	
1480	19960	49	1910	8891	89	2340	828	99.5	
1490	19944	50	1920	8813	89	2350	554	99.5	
1500	20229	51	1930	8412	90	2360	499	99.5	
1510	20170	52	1940	7999	90	2370	444	99.5	
1520	19886	53	1950	7857	91	2380	301	99.5	
1530	19668	55	1960	7521	91	2390	204	99.5	
1540	19428	56	1970	7210	92	2400	384	99.5	
1550	19075	57	1980	6931	92				
1560	19086	58	1990	6569	93				

Class Activity 8a: The Illuminated Dartboard

Suppose you projected a light onto a blank wall, and hung a dartboard so that part of it was illuminated by the projector (see our picture below). Then, blindfolded and disoriented, you threw a dart and hit a random spot on the wall. Try to answer the following two questions:

A. What is the probability that the dart hit the dartboard?
B. If you are informed that the dart landed in the illuminated region, what is the probability that it hit the dartboard?

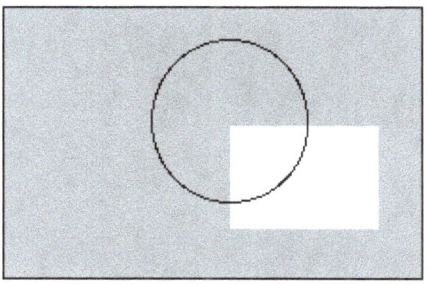

1. For question A, what did you consider the *sample space* to be? How about for question B? (It should be helpful to think of the areas of the regions involved, interpreting the probabilities using geometric reasoning.)

2. Show that you can write your answer to dartboard question B in the following form:

    $$\frac{the\ probability\ that\ the\ dart\ hit\ the\ illuminated\ part\ of\ the\ dartboard}{the\ probability\ that\ the\ dart\ hit\ somewhere\ in\ the\ illuminated\ region}$$

3. More generally, suppose A and B are two events, each having a positive probability of occurrence. Write an expression for *the probability of B, if you know A occurred,* in terms of *the probability of A* and *the probability that both A and B occur.*

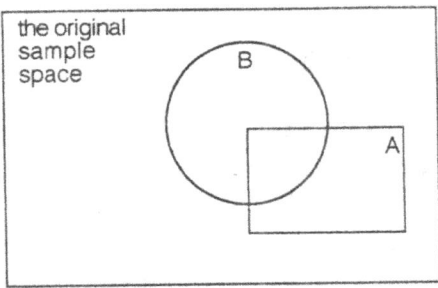

4. Show that, in order to find the probability of a compound event (or a sequence of events), you can multiply the probabilities of the individual events together, but continually taking into account all prior information as you go along.

Class Activity 8b: Three Prisoners

We are the prisoners of ideas.
Ralph Waldo Emerson

A, B, and C are three prisoners serving life sentences. (You'll need three volunteers to play these roles. You also need a volunteer to play the warden – isn't this fun?) Today the warden makes an announcement to Prisoners A, B, and C: "Because of prison overcrowding, I am going to randomly set two of you free." (Go ahead, Warden, say this.) At this point, all three prisoners must leave the room (to the detention block in the hallway), while the warden randomly names two from among them. The rest of the class should be told which two are going to be set free, but the prisoners should not know. Now the prisoners can return to the classroom (the warden's office).

Question 1: (to be answered individually by each of the three prisoners): Evaluate the probabilities of release for each of the three of you. (Each prisoner should give three responses.)

After the three prisoners justify their answers to the class's satisfaction, the following scene ensues:

Prisoner A: "Okay, Warden, who's it gonna be?"

Warden: "I can't tell you that. Not until tomorrow."

Prisoner A: "Then how about you tell me this -- I know one-a-dem other guys is gonna go free anyway, so just gimme a name. Tell me one-a-dem who's going free. I won't tell nobody nothin', and you won't be tellin' me nothin' about my own situation."

Warden (*contemplative*): "Well... okay." *The warden then whispers to A the name of a prisoner (not A) whom he has chosen to set free.*

Question 2: (to be answered by A only -- B and C don't know anything new): Again, evaluate the probabilities of release for each of the three of you.

(Question 2 will hopefully wreak havoc in the classroom. After everyone forms an opinion about this, it should be useful to perform many simulations of the activity, beginning at the point of randomly choosing which two prisoners will be set free. This will allow you to evaluate the probability through frequencies.)

Read and Study: Conditional Probability

Chance is the providence of adventurers.
Napoleon Bonaparte

The *Three Prisoners Class Activity* presents a variation on a famous problem that comes in many guises (For a few of them, see (ref 1).) One variation, known as the *Monte Hall Problem* (based on an old game show), was presented by Marilyn vos Savant (listed in the *Guinness Book of World Records Hall of Fame* for "Highest IQ") in the newspaper supplement *Parade* in 1990 (ref 2). Marilyn correctly solved the problem, but *thousands* of people, including many university professors, wrote mean-spirited letters telling her she was wrong. The professors were chastised as a "disgrace to the profession" by Leonard Gillman, former president of the Mathematical Association of America, in that association's publication, *Focus* (ref 3). On top of rushing to judgment, the professors had ignored subtleties of the game, he pointed out. The confusion arose from people's misunderstanding of *conditional probabilities*.

A **conditional probability** is a probability conditioned on certain information. In the *Class Activity*, we found that a person's evaluation of probabilities changed as new information was made available. You may have even noticed that, in contrast to the perspective of the prisoners, the audience members in the room would have perceived each of the three prisoners' probabilities of release to be either 1 or 0, because the audience had already been told who was to be released and who was to remain in prison. For another example, contrast the following two questions:

   a) If your friend rolled a fair six-sided die, what is the probability that she got a 4?

   b) If, after rolling it, she told you she got an even number, what is the probability she got a 4?

The answer to *a)* is 1/6, but the answer to *b)* is 1/3, even though both questions refer to the same roll of the die; the distinction here is that you have *different knowledge bases* on which to evaluate the probabilities.

A point often overlooked, and one you should keep in mind from now on, is that *every probability is really a conditional probability* -- *every* evaluation is based on your level of knowledge, which essentially refines your notion of the sample space; if no condition is stated explicitly, then the evaluation is implicitly based on your knowledge of the original sample space.

Sometimes your knowledge of a situation may change without affecting the probability you would assign to an event. Said in a different way, one event might occur without changing the *probability of* another event. In that case, the two events are said to be **independent**. If, on the other hand, the occurrence of one does change the probability of the other, then the events are said to be **dependent**.

We must be careful here -- people often make the mistake of thinking that dependence has to do with some tangible connection between the events, or some cause-and-effect relationship; it doesn't.

Here's an example. Suppose you rolled a fair six-sided die. Consider the following three pairs of events:

Pair 1: *rolling an even number* and *rolling a 3*

Pair 2: *rolling an even number* and *rolling a 6*

Pair 3: *rolling an even number* and *rolling a 3 or a 6*.

The events in the first pair are dependent -- the probability of rolling a *3* is 1/6, but if you rolled an even number, then the probability of its being *3* is 0. The events in the second pair are also dependent -- the probability of rolling a *6* is 1/6, but if you rolled an even number, then the probability of its being *6* is 1/3. In the third pair, the occurrence of either one of the events does limit the options for the other event, but nevertheless, the events are *independent*. *Verify this.*

When two events A and B are dependent, some people say "A is dependent on B" to reflect that the probability of A changes if you know B has occurred, or "B is dependent on A" to reflect that the probability of B changes if you know A has occurred. As it turns out, if A is dependent on B, then B must also be dependent on A. As such, you could sometimes say that an event in the past is dependent on an event in the future. (Remember, this would really be a statement about probabilities.) The modern discovery of the ruins of an ancient city is dependent on the city's having existed in the first place. (Clearly, the probability of discovering the ancient ruins of a city increases if the city actually existed to begin with.) Turning things around, the city's existence in the past is also dependent on its discovery in the future. (The probability that the city existed is '1' if in fact you have just discovered it, but you cannot be sure it existed if you have not discovered it.)

On a deeper, freaky level, the perception that we are moving through time from the past to the future might really be an illusion to begin with, due to the way our brains record information. The laws of physics (for small particles, at least) do not appear to distinguish between past and future (ref 4).

Now let's consider the following classic probability question: *A jar contains 2 red marbles and 2 white marbles. Suppose you draw one marble out of the jar, throw it away, and then draw another marble. What is the probability that the second one you draw is red?*

This simple problem is great for illustrating conditional probabilities. The probability that the second marble is red clearly depends on the result of the first draw. If the first draw was red, then there would only be 1 red left out of 3 still in the jar, so the probability the second draw is red would be 1/3. However, if the first draw was white, there would still be 2 red left out of the 3 in the jar, so the probability the second draw is red would be 2/3. We can see that knowing what happened in the first draw changes the probabilities for the second draw, so we can say that they are dependent events. A probability tree (see *Figure 3*) is a great device for keeping track of the conditional probabilities at various stages of such a problem. The probabilities of the pieces are labeled along the branches. Make sure you understand this picture.

But we have not yet found an answer to the original question: *What is the probability that the second one you draw is red?* In this situation we aren't told what the result of the first draw was. We want to know before we draw that first marble what the likelihood is that the second marble ends up being red. One way to think about this is to take a frequency view of probability: if we were to repeat this experiment many times, what proportion of the time would we end up with a result where the second draw was red. Looking at the tree, we want to know what proportion of the time would we end up taking either the "red, then red" path, or the "white, then red" path.

If we want to know the probability of a particular sequence of events, we can trace the path of the tree corresponding to the sequence of events we described, and, using the result of #4 in the Illuminated Dartboard activity, simply multiply the probabilities along that path. Thus, if we want to know the probability that both draws were red, we could multiply $\frac{2}{4} \times \frac{1}{3}$ to get $\frac{2}{12}$. If we wanted to know the probability that the first draw was white and the second was red, we would multiply $\frac{2}{4} \times \frac{2}{3}$ to get $\frac{4}{12}$. Again, using a frequency view of probability, 2 out of 12 times we'd expect to get a result of "red, then red", and 4 out of 12 times we'd expect to get a "white, then red" result. So we expect to get red on the second draw 6 out of 12 times.

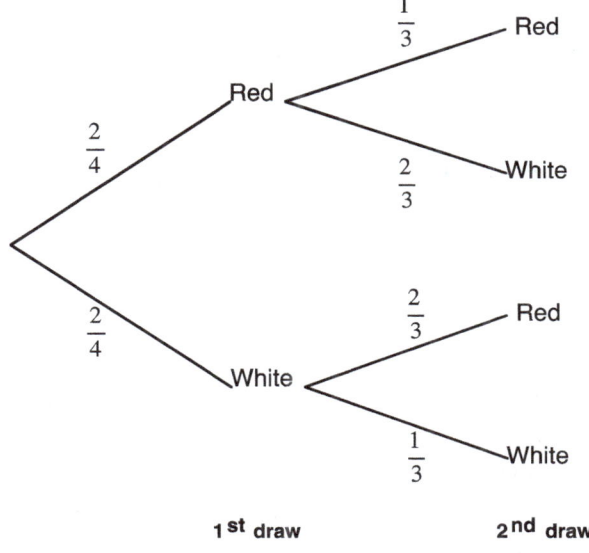

Figure 3: A Probability Tree

Homework

There are three kinds of lies: lies, damned lies, and statistics.
Benjamin Disraeli

1) You take a 5-question multiple-choice test, each question having 4 possible choices.

   a) If you randomly guess at each answer, what is your probability of getting a perfect score?
   b) What is your probability of getting the first three answers correct but missing the next two?

2) Suppose you were to spin both of the spinners pictured below

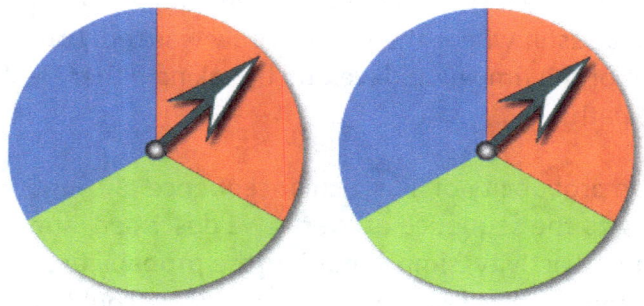

   a) What is the probability that both land red?
   b) What is the probability that one lands red and the other lands blue?
   c) If you don't see how they landed, but you are told that (at least) one landed red, what is the probability that the other one landed blue?

3) You flip a fair coin 3 times. If at least one of the flips is *heads*, what is the probability that they're all *heads*?

4) Consider the problem we explored in the Read and Study:

   A jar contains 2 red marbles and 2 white marbles. Suppose you draw one marble out of the jar, throw it away, and then draw another marble. What is the probability that the second one you draw is red?

   a) Pretend that you labeled the marbles R1, R2, W1 and W2. Use these labels to list out a sample space that consists of 12 equally likely outcomes for the experiment.
   b) Draw a large Venn diagram similar to the one in the Illuminated Dartboard Activity. Label event A as "first draw is red" and event B as "second draw is red". Place each of your 12 outcomes in the appropriate region of the Venn diagram.

c) Explain how you can use your Venn diagram to answer the question posed in the original problem: what is the probability that the second one you draw is red?
d) Make an explicit connection between each probability labeled in the probability tree in the Read and Study and the Venn diagram you made.

5. A jar contains 2 red and 3 green marbles. You draw two marbles without replacement. Find the following probabilities:

   a) They are both red.
   b) The first is red and the second is green.
   c) One is red and the other is green.
   d) The second is green, given that the first was red.
   e) The first was red, given that the second is green.

6. You roll two fair six-sided dice. Find probabilities for the following events:

   a) You roll doubles.
   b) Both dice are even.
   c) You roll doubles and both dice are even.
   d) You roll doubles, given that both dice are even.
   e) Both dice are even, given that you rolled doubles.
   f) Are "rolling doubles" and "both dice are even" independent or dependent events?

7. You interviewed 100 people. 40 of them were male. 60 of them were Packer fans. 30 of them were male Packer fans. Among these people:

   a) What is the probability that a male is a Packer fan?
   b) What is the probability that a Packer fan is a male?
   c) In this example, are the events "being a male" and "being a Packer fan" independent?

8. A Las Vegas roulette wheel is divided into 38 sectors numbered 1-36, 0, and 00. A ball dropped on the spinning wheel is equally likely to land on any of the 38 numbers. Players pick numbers; for every dollar they placed on a winning number, they win $35 profit. Multiple players may bet on the same number.

   a) One fine day, "Nevada," a lady living in Reno, decides to call in sick and drive to Las Vegas. She ends up at the same roulette table as "Tex," a gentleman from Houston who has spent the past month squandering his oil fortune. Nevada and Tex don't know each other from a monkey on a rock. Just before a spin of the wheel, Nevada places a bet on 11, because 11 is her favorite number. Tex uses a random number generator to randomly pick one of the 38 numbers; it generates the number 24, and so he places his bet on that. Are the two events *Nevada wins* and *Tex wins* dependent or independent? Explain. (Remember, the first thing to do in an instance like this is to carefully read the relevant definition.)

b) As you might expect, after awhile, Nevada and Tex got that wandering spirit, and both found themselves at yet another roulette table. This wheel, to their surprise, was unusual, in that there was only *one* section (numbered 1), and the ball *always* landed on it, so everybody always won. Of course they immediately (and simultaneously) put all their money on the number 1. Are the two events *Nevada wins* and *Tex wins* dependent or independent? Explain.

c) Nevada and Tex eventually get tired of winning and wander off in different directions, each to place one final bet. They end up at different roulette tables. Nevada places a bet on 11 at her table, and Tex places a bet on 11 at his table. Are the two events *Nevada wins* and *Tex wins* dependent or independent?

9. If you made two independent tosses of a fair coin, what is the probability of getting *TT*? What is the probability of getting *HT*? Now, suppose you tossed a coin over and over and recorded the resulting sequence of *H*'s and *T*'s. What is the probability that *TT* would occur before *HT* in the sequence? Are your answers consistent?

10. Suppose you have two coins whose tosses are independent of each other.

    a) Suppose the coins are fair. What is the probability of getting two heads? Two tails? One head and one tail?

    b) This time let's remove the assumption of fairness; for each coin, you may decide with what probability it lands heads. Show that it is *not possible* for the following three outcomes to be equally likely: two heads; two tails; one head and one tail.

    Note: Although you won't find a pair of actual coins that behave like the ones in (b), certain small particles do behave in this strange way, according to Bose-Einstein statistics (ref 8). It is as though one particle knows what the other is doing. This might not seem reasonable, but physical experiments sometimes show us that what we think *ought* to happen can be a very different thing from what *does* happen.

11. The data for this problem is taken from (ref 5); a thorough explanation of terms, methods, and data sources is provided in that article as well. Use the table below as you work through this problem. The "average number of years of life remaining" is commonly referred to as the "remaining life expectancy" (the word *expectancy* derived from our term *expected value*).

    Table values are *average number of years of life remaining, for Americans.*

Age	2002	1900-02
0	77.3	49.24
1	76.8	55.20

5	72.9	54.98
10	67.9	51.14
20	58.2	42.79
30	48.7	35.51
40	39.3	28.34
50	30.3	21.26
60	22.0	14.76
70	14.7	9.30
80	8.8	5.30
90	4.8	2.95
100	2.7	1.58

a) For the year 2002, the table says the remaining life expectancy for a 100-year old is 2.7 years. Explain, then, why the life expectancy of a newborn is not 102.7 years.

b) In 2002, the probability that a 99-year old would live at least one more year is 0.734854. Use this information, together with the information referred to in part (a), to calculate the remaining life expectancy for a 99-year old.

c) In 2002, the probability that a 97-year old would live at least one more year is 0.765053, the probability that a 98-year old would live at least one more year is 0.750113, and the probability that a 99-year old would live at least one more year is 0.734854. Find the probability that a 97-year old would live to be 99 but die before reaching 100.

d) Comparing the data from the two columns in the table, we see that people live, on average, about 28 years longer now than they did a century ago. What do you think accounts for this difference? (Consider the table carefully.)

12. In 1966, two surveys were conducted to determine the average length of stay for tourists visiting Morocco. In each survey, tourists were asked to report how long they stayed or intend to stay in the country. In one survey, the information was collected at the airport and border crossings as tourists were leaving the country. In the other survey, information was collected from tourists while they were residing at their hotels. One of these methods yielded an average of 9.0 days, and the other yielded an average of 17.8 days (ref 7). Which figure came from which survey, and how can you tell? Conduct a simulation to explain your reasoning.

Chapter Two

Counting

Class Activity 9: Seating Arrangements

Standard mathematics has recently been rendered obsolete by the discovery that for years we have been writing the numeral five backward. This has led to reevaluation of counting as a method of getting from one to ten.
Woody Allen

The class should be divided into groups of 4, 5, or 6 people (at least one group of each size, if possible). Each group should designate two people, one to be called Montague and the other to be called Capulet. Then each group should solve the following problem. (Ultimately, you should solve it theoretically, but you are also welcome to simulate the scenario and collect data.)

At a wedding reception, your group is to be seated together at a round table with exactly as many chairs as there are people in your group. The seating has been assigned randomly. What is the probability that Montague and Capulet will wind up seated next to each other?

Several different methods of solution might come out of these. Discuss them as a class. Can you extend your answers for 4, 5, and 6-person groups to n-person groups?

Next, do the same problem again, but instead of choosing two people to be Montague and Capulet, choose three people to be the Three Musketeers, and find the probability that all three will wind up seated next to each other. Try to extend your answers to the case where the Three Musketeers are sitting in n-person groups.

Read and Study: Permutations

The statistics on sanity are that one out of every four Americans is suffering from some form of mental illness. Think of your three best friends. If they're okay, then it's you.
Rita Mae Brown

Recall the recent homework problem:

A jar contains 2 red marbles and 2 white marbles. If you draw a marble, throw it away, and then draw another marble, what is the probability that the first is white and the second is red?

We were asking you to use a *probability tree* and to use the definition of a conditional probability to multiply probabilities. Now we're going to solve it a different way. There are 4 choices for the first draw, and for each of those there are 3 remaining choices for the second draw, so the total number of ways you could end up drawing two marbles out of this jar is 4×3. We illustrate this in with a *counting tree*:

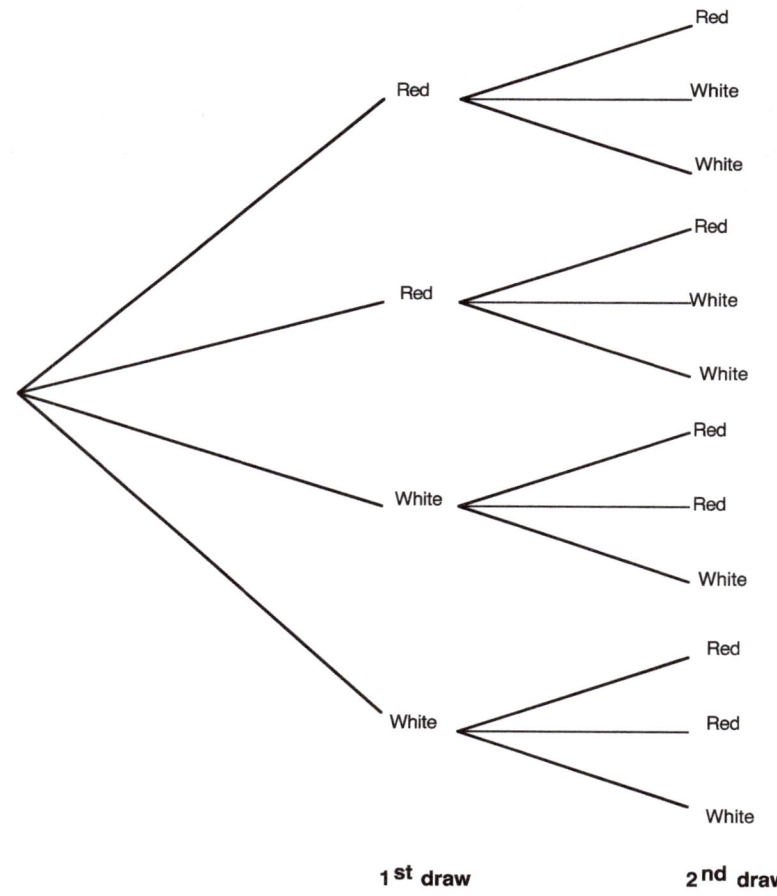

There are 2 white choices for the first draw, and for each of those there are 2 red choices for the second draw, so the total number of ways in which the first could be white and the second red is 2×2. *Trace these four paths in the counting tree.* Each of the individual outcomes is equally likely, and so the probability of drawing a white marble followed by a red marble is the ratio

$$\frac{2 \times 2}{4 \times 3}.$$

Notice that this matches up with our old answer:

$$\frac{2 \times 2}{4 \times 3} = \frac{2}{4} \times \frac{2}{3}.$$

What we have done is multiply the *counts* at the individual stages first, and then take the ratio to find the probability.

As with a probability tree, a complete sequence of events is represented on a counting tree by moving along a particular path from left to right; unlike a probability tree, rather than label how *likely* we are to move along the various branches, we require that all choices be equally likely. *What are some advantages and disadvantages of working with counting trees as opposed to probability trees, to find probabilities of sequences of events?*

A certain kind of counting situation occurs so often that we will address it specifically. First, some terminology: A **set** is a collection of distinct elements; there are various ways of writing any set, but the elements must be clearly identified. For example, the set consisting of the elements a, b, and c could be written in any of the following ways:

{a, b, c};
{b, a, c};
{a, a, b, c}.

The third listing is redundant, but still identifies the same three elements a, b, and c. A set could have an infinite number of elements, such as the set of positive integers, or it could have no elements. A set with no elements is called the *empty set* and can be denoted like this: { }. *How many empty sets are there?*

A **set permutation** is an ordered arrangement of the elements of a set. There are six permutations of the set {a, b, c}. Three of them are:

a b c

                    a c b
                    b a c

Find the other three.

A note regarding notation: We will often need to write down products such as $5 \times 4 \times 3 \times 2 \times 1$. Because this gets tiresome, we will use the shorthand notation **5!** (We read this as "5 factorial."). Of course this is not restricted to the number 5.

Homework

He who opens a school door closes a prison.
Victor Hugo

1) Consider the set {A, B, C, D}.

   a) How many permutations are there of this set?

   b) Draw the counting tree which lists all of these permutations.

   c) Suppose you had four alphabet blocks -- one with the letter A, one with B, one with C, and one with D. If you randomly arranged these blocks in a row, what is the probability that the first one would be A and the second one B?

   d) Using the setup from part c), what is the probability that the first two blocks would be A and B (in *either* order)?

2) In World War II, the German military communicated vital information through messages encrypted with a machine known as *Enigma*. Enigma looked much like a typewriter, but when you hit one letter, another letter would light up. The letters were electronically connected in pairs, so that, for instance, if you typed *A* then *Q* might light up; to decrypt this, the receiver would type *Q* on his machine, and *A* would light up. So basically, Enigma was a device which created 13 pairs of letters, swapping one letter in the pair for the other when that key was typed. The Poles were able to construct a replica of Enigma, so it seems as though, in order to decrypt an intercepted message, all they would have needed to do would be to try the various possible pairings until the message made sense. But as you'll see in this exercise, the number of such possible pairings is overwhelmingly large. Furthermore, Enigma was designed to change the pairings every time a letter was pressed, so that *Q* might light up the first time you typed *A*, but *L* might light up the second time you typed *A*. Cracking a code thus required knowing not only how the letters were initially paired, but also the pattern for changing the pairings. The machine was a brilliant creation, but there were weaknesses in its implementation,

which the Allies were able to take advantage of in a successful bid to decode the German messages. (This information has been summarized from the National Security Agency's publications ref 2 and ref 3, the two of which contain far more details regarding the construction of the machine and the history of the Allies' code-cracking methods.)

Your problem: Calculate the number of possible ways of creating 13 pairs from among 26 letters. (For example, one way would be *A-K, B-L, C-M, D-N, E-O, F-P, G-Q, H-R, I-S, J-T, U-Z, V-Y, W-X.*)

3) A delivery truck is loaded with 8 packages, one for each of the cities A, B, C, D, E, F, G, and H. The cities are arranged alphabetically in a circle, as pictured below, centered at the delivery hub. It takes 1 hour for the truck to drive from the hub to any of the 8 cities, 1 hour to drive from any city to an adjacent city, and 2 hours to drive from any city to a nonadjacent city (by passing through the hub). Assume that it takes no time at all to actually deliver a package once the truck has reached a destination.

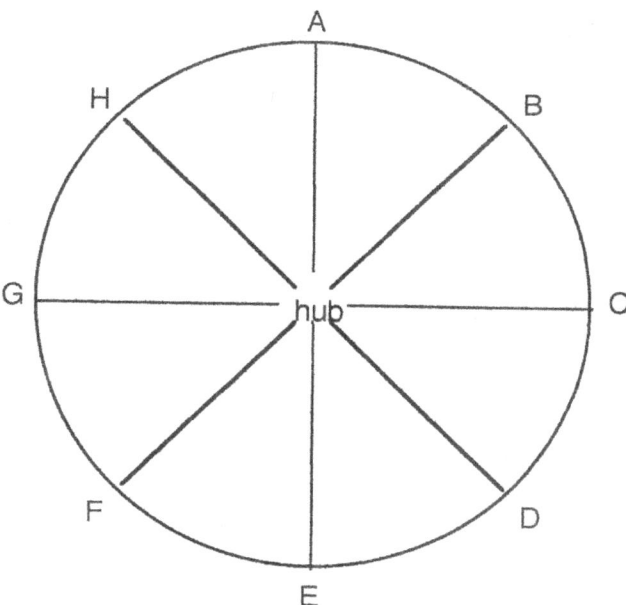

Before making his deliveries, the driver randomly chooses the order in which he will visit the cities. (One possibility would be C,B,F,E,H,D,A,G.) Then he begins his route at noon.

a) What is the probability that the truck will have completed its deliveries and returned to the hub before 9:30 that night?

b) What is the probability that the first 3 cities to receive delivery will be A, B, and C (in any order)? In part b), we no longer care whether the truck returns by 9:30.

Note: This is related to the famous *traveling salesman problem* (ref 3), a topic highly relevant to many business operations requiring efficiency of vehicle traffic, delivery time, computational speed, etc. The numbers for problems of this type go out of control surprisingly quickly. If a delivery truck had packages to deliver to just 15 locations, then the number of different possible delivery routes is bigger than a trillion! *Verify this.*

4) Suppose you have a group of 25 people (such as math class). Estimate the probability that there are no birthdays in common among the people in the group (in other words, that everyone in the group has a different birthday). What is the probability that at least 2 of the people in the group share a birthday?

5) You have 4 grey marbles and 12 white marbles, to be randomly placed in a square box divided into 16 compartments; two possible arrangements are illustrated below. What is the probability that the 4 grey marbles will form a 2-by-2 square (an example of which is illustrated in the picture on the left)?

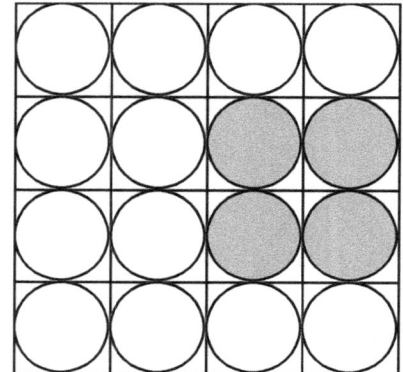 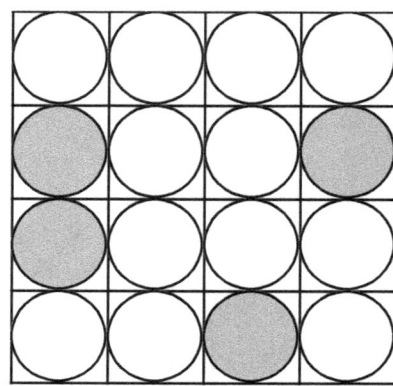

Note: The result would seem to tell us that nicely-ordered arrangements are unlikely to occur randomly. But this is deceptive. In most cases, the kinds of arrangements that we *perceive* as nicely ordered are very few compared to the kinds of arrangements that we perceive as haphazard. It is a fundamental principle of physics that, within a closed system (one that exchanges no energy with outside systems), the level of disorder increases over time (ref 4). But if you put energy into a system, then the level of order can increase within that system. For example, if you magnetized the grey marbles in the problem above, and jiggled the marbles around to randomize them, then almost surely the grey ones would end up stuck together. Here's one you can try at home: Boil some water (that puts energy into the system) and over saturate the water with sugar. Then let it cool, pour it into a glass, and let it sit for a week or two. The cool water will not hold as much dissolved sugar as the hot water, so some of the sugar molecules come out of the solution. These molecules tend to stick together (like weak magnets) when they bump into each other. They will end up forming large crystals. So order *can* randomly arise from disorder!

Class Activity 10: M I S S I S S I P P I

Each group should take 11 index cards or sheets of paper. On one of them, write an *M*; on four of them, write an *I*; on four of them, write an *S*; on two of them, write a *P*. If you randomly arrange these cards in a row, what is the probability that you will have created the word *MISSISSIPPI*? Work with your groups to solve this, and then share your approaches with the class.

Read and Study: Multiset Permutations

In the last section, we learned about *set permutations*. In this section, we will explore a related idea, **multiset permutations** (ref 1). Like a set, a multiset is a collection of elements, and it doesn't matter in what order those elements are listed; but unlike a set, it does matter how many times each element appears. We illustrate this with the following examples:

{a, a, b, c} is the same set as {a, b, c}, and is usually just written as {a, b, c};
{a, a, b, c} and {a, b, c} are not the same multisets.

A multiset permutation is an ordered arrangement of the elements of a multiset, where the repeated elements are regarded as indistinguishable from each other. There are twelve different permutations of the multiset {a, a, b, c}. Three of them are:

a a b c
a b a c
a c a b

Find the other nine and list them here:

When the multiset is a collection of letters, a permutation of that multiset is really just the same thing as a word that uses all of those letters (where by "word" we are also including meaningless things such as "abac".) So one approach to Class Activity 8 could involve figuring out how many permutations there are of the multiset {M, I, S, S, I, S, S, I, P, P, I}.

Consider the multiset {Y, Y, Y, Y, N, N, N}; this multiset of course has many permutations (you will count them in *Homework # 1*.) One of them is

Y N Y N Y Y N.

Let's write this more carefully, identifying the positions below each letter:

Y	N	Y	N	Y	Y	N
1st	2nd	3rd	4th	5th	6th	7th

A way of identifying this permutation is to state which 4 positions are occupied by Y's (the 1st, 3rd, 5th, and 6th) and which 3 positions are occupied by N's (the 2nd, 4th, and 7th). In the homework we will ask you to work through a series of exercises that will allow you to develop your understanding about multi-set permutations and a way to compute how many permutations a multi-set will have.

Homework

1. Consider the following multiset: {Y, Y, N}

    a) How many permutations are there of this multiset? List them all.
    b) Now pretend that the two Ys in the multiset are distinguishable, by labeling them Y_1 and Y_2. Then the multiset would become a set $\{Y_1, Y_2, N\}$. How many permutations are there of this set? List them all.
    c) Match each set permutation in part b) with the corresponding multiset permutation in part a). How many set permutations did each multiset permutation get matched with?

2. Consider the following multiset: {Y, Y, Y, N}.

    a) How many permutations are there of this multiset? List them all.
    b) Now pretend that the three Ys in the multiset are distinguishable, by labeling them Y_1, Y_2 and Y_3. Then the multiset would become a set $\{Y_1, Y_2, Y_3, N\}$. How many permutations are there of this set? List them all.
    c) Match each set permutation in part b) with the corresponding multiset permutation in part a). How many set permutations did each multiset permutation get matched with?

3. Consider the following multiset: {Y, Y, N, N}.

    a) How many permutations are there of this multiset? List them all.
    b) Now pretend that the two Ys in the multiset are distinguishable, by labeling them Y_1 and Y_2, and pretend that the two Ns are distinguishable, by labeling them N_1 and N_2. Then the multiset would become a set $\{Y_1, Y_2, N_1, N_2\}$. How many permutations are there of this set? List them all.
    c) Match each set permutation in part b) with the corresponding multiset permutation in part a). How many set permutations did each multiset permutation get matched with?

4. a) Without listing them, find the number of permutations of the multiset {Y, Y, Y, Y, N, N, N}, and explain how you know that your answer is correct.

    b) Consider the set {A, B, C, D, E, F, G}. How many different subsets of four elements could be chosen from this set? Why is this part b) of problem 4?

5. How many permutations are there of the multiset that consists of r copies of Y and k copies of N?

6. In the picture below, a fire truck is located at the intersection of 1st St. and Ave A, and a fire rages at the intersection of 5th St. and Ave F. The fire truck travels 1 block per minute.

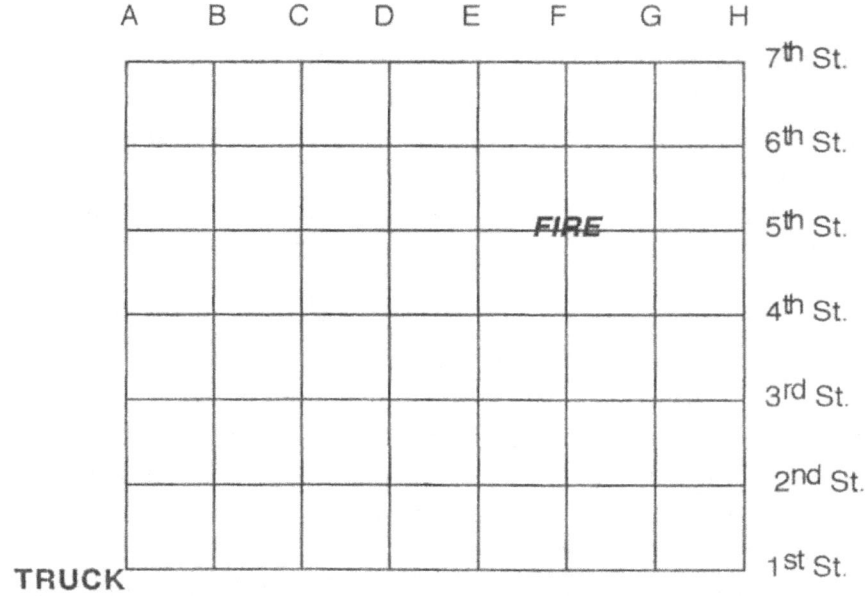

a) By how many different paths could the fire truck reach the fire in 9 minutes?

b) Suppose the fire truck has successfully reached the intersection of 3rd St. and Ave D in 5 minutes. But the smoke is now so thick that the driver can't see where the fire is coming from, so beginning with that intersection, he randomly decides whether to turn right, turn left, or go straight ahead. What is the probability that the truck will reach the fire in the remaining 4 minutes?

Class Activity 11: Pascal's Triangle

A **combination** of a set is a subset of that set.   Consider a set of 5 objects: {A,B,C,D,E}

- How many subsets can you form that contain all 5 of these elements? List them all.

- How many subsets can you form that contain exactly 4 of these elements? List them all.

- How many subsets can you form that contain exactly 3 of these elements? List them all.

- How many subsets can you form that contain exactly 2 of these elements? List them all.

- How many subsets can you form that contain only 1 of these elements? List them all.

- How many subsets can you form that contain none of these elements? List them all.

The *number* of combinations of size r taken from a set of size n is denoted $_nC_r$.

The values for $_nC_r$ for increasing values of n and r be arranged into what is called Pascal's Triangle:

$$_0C_0$$

$$_1C_0 \quad _1C_1$$

$$_2C_0 \quad _2C_1 \quad _2C_2$$

$$_3C_0 \quad _3C_1 \quad _3C_2 \quad _3C_3$$

$$_4C_0 \quad _4C_1 \quad _4C_2 \quad _4C_3 \quad _4C_4$$

.
.
.

Fill in the values for the first several rows of Pascal's Triangle. What patterns do you see? Make as many conjectures as you can and try to prove that your conjectures are correct and make sense.

Read and Study: Combinations

The concept of a combination, or a subset of a set, is a crucial one in counting, and we will rely on this concept, in the remainder of this book. As such, it is convenient to develop formulas for computing the number of combinations of a given size. One handy tool for small set sizes is Pascal's Triangle, but this is not convenient for larger numbers.

In the two previous sections, we have asked you to come up with ways to calculate the number of permutations of a set, as well as the number of permutations of a multiset. The good news is that we can make explicit connections between a permutation and a combination. And we can also make an explicit connection between a multiset permutation and a combination. So that will give us two different (but of course related) ways of thinking of combinations, and two different (but of course equivalent) ways of deriving a formula for $_nC_r$.

First we will ask you to make an explicit connection between combinations and multiset permutations. For example, let's consider $_5C_3$. This is the number of combinations of size 3 taken from a set of size 5. To compute this number, we can take a set of size 5, say {A,B,C,D,E}, and list out all of its subsets of size 3. *Do this now. How many subsets are there?* Another way of thinking about forming subsets of size 3 would be to consider each element in sequence, and deciding whether or not to include this element in our subset. First we can consider A. Should we include it in our subset or not? We have a decision: Yes or No. Then we can move to element B. Should we include this element or not? Then we can move on to C, then D, and finally E.

We can record our decisions as a sequence of Yes's and No's. For example, a sequence of Yes, No, Yes, No, Yes (or YNYNY for short) would correspond to choosing A, not choosing B, choosing C, not choosing D, and choosing E, resulting in the subset {A, C, E}. Similarly, the sequence NNYYY results in the subset {C,D,E}. So there is a one-to-one correspondence between sequences of three Y's and two N's, and subsets of a set with 5 elements. In other words, there is a one-to-one correspondence between the permutations of the multi-set {Y,Y,Y,N,N} and the combinations of the set {A,B,C,D,E}. *Complete this one-to-one correspondence: for each subset of size 3 for {A,B,C,D,E}, write its corresponding permutation of {Y,Y,Y,N,N}.*

So a combination is equivalent to a multiset permutation with only two categories of elements: the ones you are choosing and the ones you are not. In the previous section's homework, we asked you to develop a formula to compute the number of permutations of the multiset that consists of r copies of Y and k copies of N. This same formula can then be used to compute the number of combinations of size r taken from a set of size $r + k$. We will ask you to do this in the homework for this section.

The inter-connectedness between the concepts of permutations, multi-set permutations and combinations manifests itself in there often being multiple ways of thinking about probability problems involving counting. Here is a classic example:

A jar contains 10 red marbles and 20 white marbles. You take 10 marbles out of the jar, without replacement. What is the probability that all 10 marbles will be red?

Spend some time now on this problem and see if you can come up with an answer. Then see if you can find another way to find an answer. We'll put some different ways we thought of at the end of this section for you to consider.

Homework:

1. Suppose you have 20 books, and you want to line up 5 of them on a bookshelf. How many different ways could you do this? Explain your reasoning.

2. Suppose you have 20 books, and want to give 5 of them to your friend. How many different ways could you do this? Explain your reasoning.

3. In how many ways can a group of 12 employees be divided up so that five work the early shift, four work the second shift, and three work the late shift? Explain your reasoning.

4. In the read and study, we presented the following scenario: A jar contains 10 red marbles and 20 white marbles. You take 10 marbles out of the jar, without replacement. Below are two more questions about this scenario. For each question, try to find more than one way to think about the problem and write the solution.
   a) What is the probability that all 10 marbles will be white?
   b) What is the probability that 5 marbles will be red and 5 will be white?

5. A pizzeria offers a "build-your-own" option for pizzas. They have 6 different toppings available that you can choose, and you can choose as many toppings as you like. You can choose no toppings, (which would be a plain cheese pizza), or choose all 6 toppings (the "works"), or any number of toppings in between. How many different "build your own" pizzas can be made? Explain your reasoning.

6. Here are some conjectures that come from patterns you may have noticed in Pascal's Triangle. For each, make an argument for why each conjecture is true and makes sense.
   a) $_nC_0 = 1$
   b) $_nC_n = 1$
   c) $_nC_1 = n$
   d) $_nC_{n-1} = n$
   e) $_nC_r = {_nC_{n-r}}$
   f) $_nC_r = {_{n-1}C_{r-1}} + {_{n-1}C_r}$
   g) $_nC_0 + {_nC_1} + {_nC_2} + \ldots + {_nC_{n-1}} + {_nC_n} = 2^n$

7. Let $_nC_r$ denote the number of combinations (subsets) of size r taken from a set of size n, and let $_nP_r$ denote the number of permutations of r elements taken from a set of size n. What is the precise numerical relationship between $_nC_r$ and $_nP_r$. Explain why this numerical relationship makes sense.

8. Use the formula you developed to compute the number of permutations of the multiset that consists of r copies of Y and k copies of N to compute the number of combinations of size r taken from a set of size $r + k$.

9. Wisconsin offers the Badger Five Lottery. The game costs $1 to play. You choose 5 different numbers from 1 to 31. Then 5 "winning numbers" are drawn. If all five of your numbers match the winning numbers, in any order, then you win the jackpot, which starts at $10,000. If you match 4 of the 5 winning numbers, you win $50. If you match 3 of the 5 winning numbers, you win $2. If you match two of the winning numbers, you win $1. Assuming the minimum jackpot of $10,000, what is the expected value of this game?

10. The Badger Five Lottery's jackpot starts at $10,000 and increases each day that it is not won. Since 2003, the largest Jackpot in Badger Five has been $288,000. How big does the jackpot need to be in order for the Badger Five Lottery to be a fair game?

11. Most pairs of cities in America are not connected by a direct plane flight, even when both cities are served by the same airline. Why not? What do the airlines do instead? Look at the route map of a major airline, and support your answer with some specific facts.

12. Twelve Card Poker is played with a 12-card deck (just the twos, threes, and fours, for example). Each player is dealt 3 cards. The ranking of the hands is determined by how rare the hand is. The more unlikely it is to be dealt a hand, the better that hand is. Consider the following types of hands: Flush (all three cards are the same suit), Straight (one card of each value, but not all the same suit), Three-of-a-kind (all three cards are the same value), and Pair (two cards are the same value, and the other card is a different value). Rank these hands based on probability.

Answers to Selected Problems:

Below are three variations on the answer to the problem we posed in the Read and Study, each achieved by thinking about the problem a different way. Make sure you understand each form of the solution and how it is interpreting the problem.

$$\frac{10}{30} \cdot \frac{9}{29} \cdot \frac{8}{28} \cdot \frac{7}{27} \cdot \frac{6}{26} \cdot \frac{5}{25} \cdot \frac{4}{24} \cdot \frac{3}{23} \cdot \frac{2}{22} \cdot \frac{1}{21}$$

$$\frac{10 \cdot 9 \cdot 8 \cdot 7 \cdot 6 \cdot 5 \cdot 4 \cdot 3 \cdot 2 \cdot 1}{30 \cdot 29 \cdot 28 \cdot 27 \cdot 26 \cdot 25 \cdot 24 \cdot 23 \cdot 22 \cdot 21}$$

$$\frac{1}{\left(\dfrac{30!}{10!\,20!}\right)}$$

Here are three ways to think about problem 4a:

$$\frac{20}{30} \cdot \frac{19}{29} \cdot \frac{18}{28} \cdot \frac{17}{27} \cdot \frac{16}{26} \cdot \frac{15}{25} \cdot \frac{14}{24} \cdot \frac{13}{23} \cdot \frac{12}{22} \cdot \frac{11}{21}$$

$$\frac{20 \cdot 19 \cdot 18 \cdot 17 \cdot 16 \cdot 15 \cdot 14 \cdot 13 \cdot 12 \cdot 11}{30 \cdot 29 \cdot 28 \cdot 27 \cdot 26 \cdot 25 \cdot 24 \cdot 23 \cdot 22 \cdot 21}$$

$$\frac{\left(\dfrac{20!}{10!\,10!}\right)}{\left(\dfrac{30!}{10!\,20!}\right)}$$

Here are two ways to think about problem 4b:

$$\left(\dfrac{10!}{5!\,5!}\right) \cdot \frac{(10 \cdot 9 \cdot 8 \cdot 7 \cdot 6) \cdot (20 \cdot 19 \cdot 18 \cdot 17 \cdot 16)}{30 \cdot 29 \cdot 28 \cdot 27 \cdot 26 \cdot 25 \cdot 24 \cdot 23 \cdot 22 \cdot 21}$$

$$\frac{\left(\dfrac{10!}{5!\,5!}\right) \cdot \left(\dfrac{20!}{5!\,15!}\right)}{\left(\dfrac{30!}{10!\,20!}\right)}$$

Class Activity 12: Gimme Five!

I. Five fair coins are tossed. Let X be the number of *head*s that appear. In your group, repeat this experiment 100 times and make a histogram of the experimental probability distribution for the variable X. Then determine the theoretical probability distribution for X and make a histogram of this distribution.

II. Five fair dice are tossed. Let Y be the number of *ones* that appear. In your group, repeat this experiment 100 times and make a histogram of the experimental probability distribution for the variable Y. Then determine the theoretical probability distribution for X and make a histogram of this distribution.

Read and Study: The Binomial Distribution

Let's consider two probability questions, both of the same basic type:

   A. If a couple has 8 children, what is the probability that 3 are girls and 5 are boys?

   B. Suppose you took a multiple-choice exam comprising 8 questions, each question having 4 choices, but you had no idea how to answer any of the questions, and so just randomly guessed on all of them. What is the probability that you would get exactly 3 of them correct?

How are these questions similar? Why did we claim they are of the same type? *Take a minute to think about this.*

You may have noticed that in each scenario, there are several independent trials of the same procedure. (In each case there were 8 trials. In the first there were 8 children being had, in the second there were 8 questions being answered). Each trial is thought of having only two possible outcomes. (In the first, each child could be either a boy or girl. In the second, each question could be answered either correctly or incorrectly). Furthermore, the probability governing which of the two outcomes happens in each trial remains the same for every trial.

Let's first consider scenario A. But instead of focusing on just one particular combination of girls and boys, let's consider all of the different possible combination of girls and boys. If we let the random variable X represent the number of girls in a family with 8 children, then X could take values anywhere from 0 girls up to 8 girls. If we draw a histogram representing the probability distribution for the number of girls, we would get a graph like this:

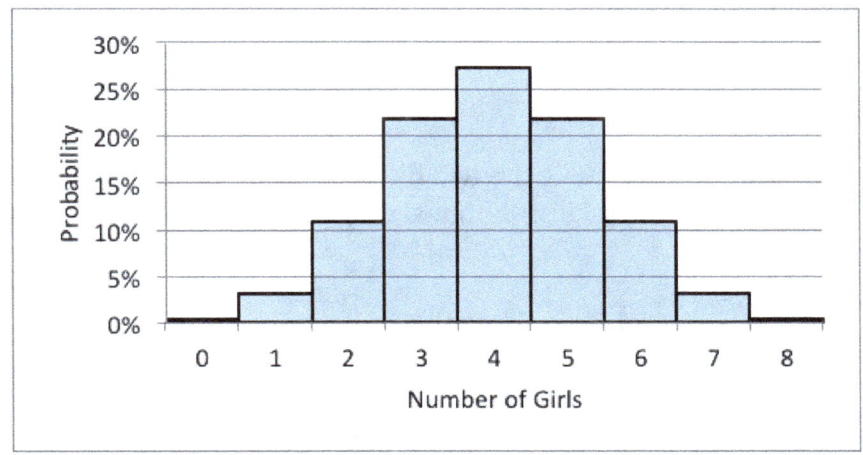

In the homework we will ask you to verify the probabilities used to make this graph.

Similarly, we can make probability distribution for the number of questions answered correctly on that 8-question multiple-choice exam in scenario B above. In the homework we will also ask you to verify the probabilities used to make this graph.

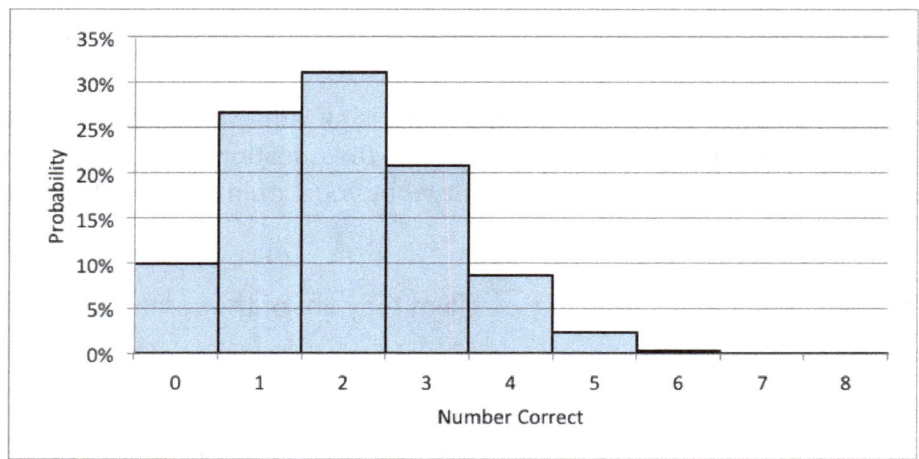

Both of the distributions above are examples of a what's known as a **binomial distribution**. Suppose there is random experiment that consists of several independent trials, each of which has two (hence, the prefix bi) possible outcomes. In general, let's say the two possible outcomes are called "success" and "failure". The trials are independent means that the probability governing those outcomes remains the same for every trial. Then the distribution of the number of trials that result in "success" is called a binomial distribution.

As we will see, binomial distributions turn out to be very common and so it will prove very useful to be able to recognize them and know how to compute binomial probabilities.

Homework:

1. Verify the probabilities used to make the two histograms in the read and study. If both of these histograms are of binomial distributions, why aren't these two histograms identical to each other? Explain.

2. Suppose you toss a fair coin 10 times.
    a) What is the probability that all 10 tosses will be *heads*?
    b) What is the probability that the first 4 tosses will be *heads* and the last 6 will be *tails*?
    c) What is the probability that the second, third, fifth, and seventh tosses will be *heads* and the other 6 will be *tails*?
    d) What is the probability that exactly 4 of the tosses will be *heads*?

3. Suppose you toss a fair six-sided die 10 times.
   a) What is the probability that all 10 tosses will be *fives*?
   b) What is the probability that the first 4 tosses will be *fives* and the last 6 will not?
   c) What is the probability that the second, third, fifth, and seventh tosses will be *fives* and the other 6 will not?
   d) What is the probability that exactly 4 of the tosses will be *fives*?

4. Suppose there is random experiment that consists of n independent trials, each of which has two possible outcomes: "success" and "failure". Let p be the probability that any one trial results in "success". Write a formula for the probability that in n trials, there will be exactly k successes.

5. If you expand the binomial expression $(x+y)^8$, what is the coefficient of the x^3y^5 term? Explain how you figured it out. If you expand the trinomial expression $(x+y+z)^8$, what is the coefficient of the x^3y^4z term? Explain how you figured it out.

6. In many sports, playoff winners are determined by a series of games. For example, in baseball, the World Series winner is determined by a seven game series. The team that wins the most games out of seven is the winner. One reason for playing more than one game is to reduce the chance that the inferior team will win just by getting lucky. Suppose there are two teams, the Argonauts and the Bean-eaters, and further suppose that the Argonauts are the superior team and will win 75% of the time that these two teams play each other. So if the championship consisted of just a single game, the probability that the Argonauts would win the championship would be 75%. Consider three possible series formats: a three game series, a five game series and a seven game series. For each of these formats, determine the probability that the Argonauts end up winning the series.

   Note: In practice, if one team wins enough games in a series so that the other team cannot possibly win more games than the other, they stop playing and do not complete the remaining games. For example, if in a seven game series, team A wins game 1, game 2, game 4 and game 5, they will not play the final two games since team A is certain to win more games than team B, regardless of the outcome of the last two games. However, to simply your analysis, you may assume that all of the games in the series would be played.

7. Shown below is the official archery target. The entire target has a diameter of 80cm, and is divided into five concentric color zones. Each color zone has a width of 8 cm, and is divided by a thin line into two zones 4 cm wide, making ten scoring zones of equal width when measured from the center. Proceeding from outside towards the center, the scoring zones are worth 1 and 2 points (white), 3 and 4 points (black), 5 and 6 points (blue), 7 and 8 points (red), and 9 and 10 points (yellow). In addition, the 10 point yellow zone is further divided into two 2 cm-wide zones, creating a special "inner ten" zone at the center of the target.

   Suppose you shoot 20 arrows at this target. Not being a skilled archer, only 10 of the arrows wind up hitting somewhere on the target.

a. What is the probability that at least 3 of your arrows land in a yellow zone? Explain your reasoning.

b. What is your expected point total?

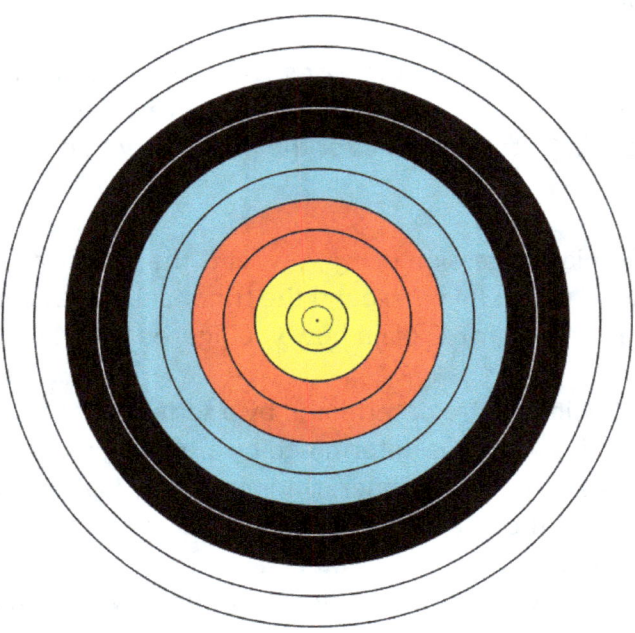

Image by Alberto Barbati, Wikimedia Commons, https://creativecommons.org/licenses/by-sa/2.5/deed.en

8. It's been a while since we've asked you about the Common Core. Read the three Common Core State Standards 7.SP.C.8.A, 7.SP.C.8.B and 7.SP.C.8.C in grade 7 under "Find probabilities of compound events using organized lists, tables, tree diagrams, and simulation". For each topic of competency mentioned, specify an example or activity (from the Class Activity, Read and Study, or Homework) from this or other sections of this text that address these standards. Also identify any parts of these standards have not (yet) been addressed in this book.

Class Activity 13a: All That and a Bag of Chips

Materials needed: Three bags containing counters.

> **Bag A** contains ninety-nine counters, numbered 1 to 99.
> **Bag B** contains one counter labeled 1 and one counter labeled 99.
> **Bag C** has eight counters labeled 0, one counter labeled 200, and one labeled 300.

Your group is going to be drawing fifteen random samples from each of the bags, where each sample consists of 10 draws *with replacement* from the bag. You will then make a line plot or histogram in each case that displays the **means** of your fifteen samples. Stop and be sure that everyone in your group understands the plan.

What is your prediction about the shape of each graph?

Now go ahead and do it. What happens in each case?

Class Activity 13b: A Dicey Situation

Rolling a die and recording its value is a classic example of a random experiment. Each roll of a die is a single observation selected at random from a population of all possible and potential rolls of the die.

Describe what you think the theoretical distribution of single observations taken from this population should be like. Make a graph of this theoretical distribution.

In this activity, you will simulate taking many samples of various sizes from this population you described above and recording the sample means.

- First, take samples of size 2. Roll two dice 30 times and record the mean value of the two dice. Make a graph of this distribution.

- Now take samples of size 4. Roll four dice 30 times and record the mean value of the four dice. Make a graph of this distribution.

- Now take samples of size 8. Roll eight dice 30 times and record the mean value of the eight dice. Make a graph of this distribution.

Compare your distributions for samples of size 2, 4, and 8. In what ways are the distributions similar? In what ways are they different? Describe the behavior of the distribution of sample means as the sample size increases.

Read and Study: Sampling Distributions

Back in the first two sections of this text we discussed the nature of statistics, in which we take a sample from some population in order to make inferences about some parameter of the population. From the data in our sample, we can calculate a statistic, but in order to make valid conclusions about our population based on our sample, we need to know how reliable our statistic really is. Aside from sample bias and confounding variables, our statistics might not be reliable due simply to random chance. Perhaps just by random chance our sample was not very representative of the population.

In the third section of the text, we explored randomness further, and saw that random processes can have quite a bit of unexpected variation. Naturally we should expect variation in outcomes, but just *how* the outcomes vary can also vary. For example, we saw that it is not uncommon to flip 6 or more heads in a row. If a sequence of 6 heads in a row were the only 6 flips in our sample, we might suspect that we have an unfair coin, when in fact it could just be that we by chance got an unexpected sample.

Now we are ready to further investigate precisely the variation that can occur between different random samples taken from the same population. The general idea is this: if we understand what happens when you repeatedly take many different samples from the same population (what kinds of samples are we likely to get, and how much variation is there from sample to sample), then we can determine what kinds of sample results would rarely happen just by chance alone. To develop this understanding, we need to learn more about theoretical distributions of samples.

It is very important in this discussion to keep in mind that we are talking about two different kinds of distributions here. To start with, we have the theoretical distribution of the population itself. We'll call this a population distribution. Then we have the distribution of some statistic calculated from different samples taken from this population. We'll call this a **sampling distribution.** In statistics, we usually don't actually know everything about the population distribution; that's why we are taking samples in the first place. But in order to understand what happens with sampling, we will sometimes pretend we know what the population distribution looks like, and then explore how the samples taken from this population can vary.

So in the class activities you had the experience of simulating taking many samples from the same underlying distribution. In *All That and a Bag of Chips,* there were three different populations, represented by the three bags. We knew exactly what these populations were like. Even though there were just a few chips in each bag, since we were always taking only one chip out and replacing it before taking another, we could think of each bag representing a very large or even infinite population. *If you haven't already, draw histograms representing the theoretical distributions for each of these three populations. Also compute the mean and standard deviation for each population distribution.* Keep in mind that these are the theoretical *population* distributions, and the mean and the standard deviation for each population are parameters (which we can denote μ and σ, respectively). Note that in the activity, we actually knew what the population parameters were, unlike in a real statistical study. But this will allow us to understand how samples behave, so we can then later apply this knowledge in cases when we don't know the population parameters.

For each population you took repeated samples, and you graphed the distribution of the means of those samples. That created a sampling distribution. In *All That and a Bag of Chips,* you were always taking samples of size 10 and recording the sample means; then you graphed the distribution of those sample means. This graph was your experimental estimate of the theoretical sampling distribution for sample size 10 taken from that population. Hopefully you observed that regardless of what the population distribution looked like, the sampling distributions all tended to be bell-shaped. In each case, the mean of the sampling distribution seemed to be the same as the mean of the population, but the standard deviation for the sampling distribution was much smaller.

In *A Dicey Situation,* there was only one population, the theoretical population of all rolls of a fair six-sided die. We also know exactly what this theoretical population looks like. In the class activity, we asked you to draw a histogram for this population distribution. *If you haven't yet, also compute the mean and standard deviation for this population distribution.* Once again, keep in mind since this is a theoretical population distribution, its mean μ and standard deviation σ are parameters. Then you took repeated samples from this population. First you took samples of size 2, and graphed the distribution of the means of these samples. This graph was your experimental estimate of the theoretical sampling distribution for sample size 2 taken from the population of all rolls of a fair die. Then you did this again, this time taking larger samples, and then even larger samples. Hopefully what you observed is that as the sample size increased, the sampling distributions got progressively more bell-shaped, and the distributions got increasingly clustered close to the population mean. That is, as the sample size increased, the standard deviation of the sampling distribution decreased.

What you were observing in the class activities are summarized by two of the most fundamental theorems of probability and statistics. The first is the **law of large numbers**. Its exact statement and its proof are pretty difficult, and too advanced to include in this book, but essentially it tells us that, under the right conditions, the sample mean is likely to be close to the population mean, provided the sample size is a *large number*.

The histograms in *Figures 1-3* below illustrate the idea; each represents the predicted distribution of the mean outcome of several rolls of a fair die. *Figure 1* is for 2 rolls (picture that we rolled a pair of dice and took the average, then we did it again, and again…), *Figure 2* is for 4 rolls (imagine that we rolled 4 dice and took their average, then we did it again and again…), and *Figure 3* is for 8 rolls. In other words, these are theoretical sampling distributions for sample sizes of 2, 4, and 8 respectively. *Take the time to understand these graphs.* Remember these graphs are histograms, which means that it is the *areas* of the bars that represent how frequently the various outcomes occurred. *What exactly is graphed on the x-axis? The y-axis?* Notice that as the number of rolls (that is, the sample size) gets larger, the sample means cluster more closely to 3.5 (the theoretical mean of any roll of the die. This is essentially what the law of large numbers says will happen.

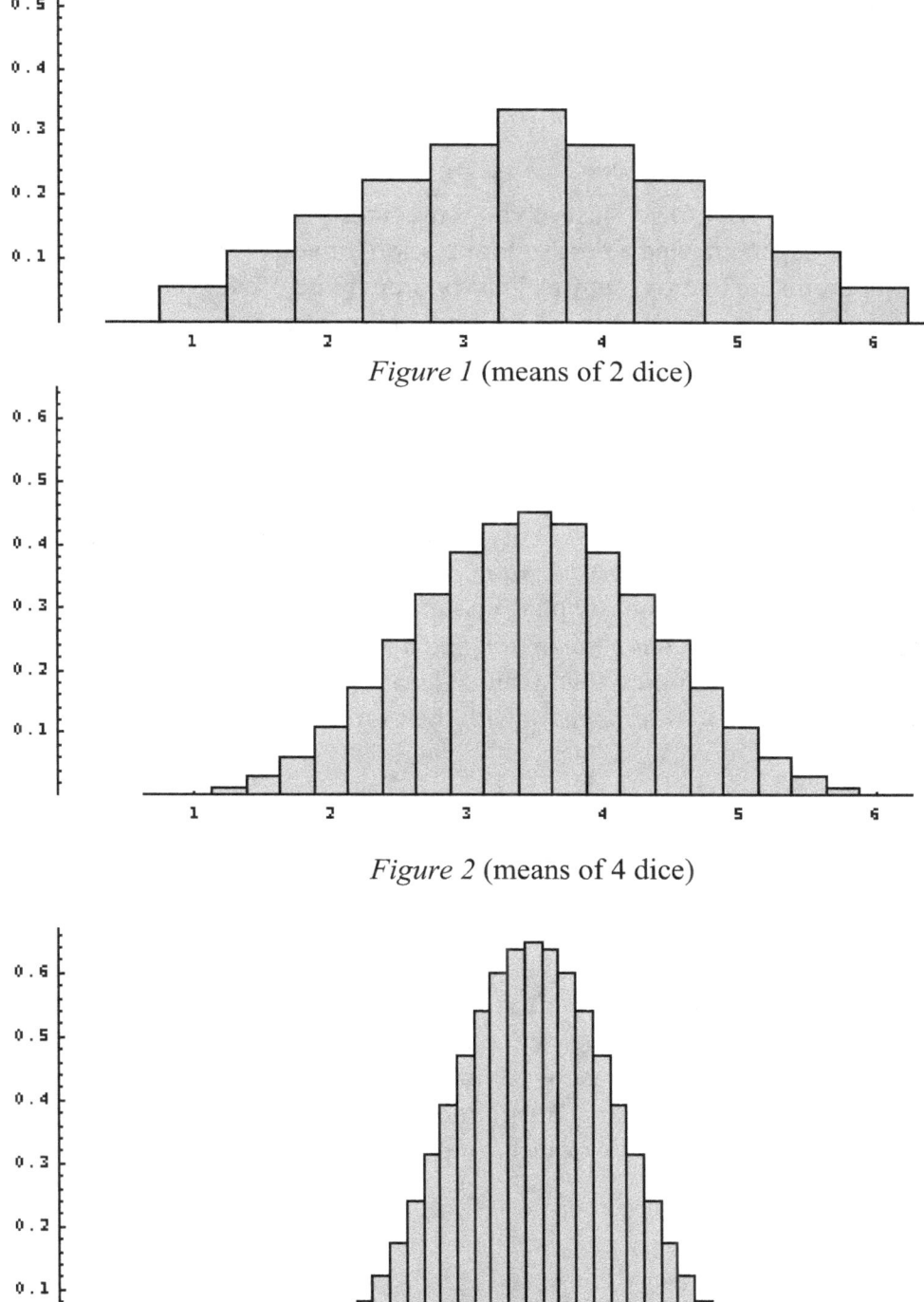

Figure 1 (means of 2 dice)

Figure 2 (means of 4 dice)

Figure 3 (means of 8 dice)

Notice in Figures 1-3 how, as the sample sizes progress from 2 to 4 to 8, the sample means cluster progressively closer to the theoretical mean in an increasingly bell-shaped distribution. A result known as the **central limit theorem** tells us that under certain conditions, as the sample size grows larger, the distribution of the sample means becomes narrower and more bell shaped, approaching what is called a *normal distribution*.

Normal distributions fit under a smooth bell-shaped curve given by an algebraic formula $f(x) = \frac{1}{\sqrt{2\pi\sigma^2}} e^{-\frac{(x-\mu)^2}{2\sigma^2}}$. Don't worry; we won't be asking you to remember or to use this formula in this text. We wanted to show it to you just so you remember that the concept of a normal distribution is well-defined, and having a precise formulation for what it means to be bell-shaped allows for theorems such as the central limit theorem to be proven. One thing we can notice is that this formula for a normal distribution is a function of only the mean and standard deviation for the distribution. So basically, if we specify a mean μ and a standard deviation σ for the population, we can know exactly what the normal distribution with those parameters looks like by graphing and analyzing this function.

Moreover, the central limit theorem also gives a value for the standard deviation of the normal distribution that the theoretical sampling distribution approaches. Suppose our underlying population has a mean of μ and a standard deviation σ, and we are taking samples of size n from this population. As n increases, the sampling distribution of the sample means will approach a normal distribution that has a mean of μ and a standard deviation of $\sigma/\sqrt{n}$. In the homework, you will be asked to compute the mean and standard deviation for the sampling distribution in Figure 1 to see how they compare with the predictions of the central limit theorem.

In a previous section we explored randomness, and you may have been surprised to see the amount of variation that can occur with random processes. However, what we have seen in this section is that even though this variability can make it difficult to predict the outcome of a single experiment or of a small sample, in the long term, random processes behave very predictably. The law of large numbers and the central limit theorem can be very reassuring in that way. They also give us something definite that we can work with. Instead of being faced only with distributions that are always changing and unpredictable, if we look to the long run, namely by considering the sample means as the sample size gets larger and larger, we start to see regularity and predictability. The normal distribution and the central limit theorem allow us to make precise predictions about how we expect samples to behave.

Homework:

Note: 1 and 2 are labor intensive to do by hand. It is recommended you use a computer spreadsheet program to do the calculations.

1. In *All That and a Bag of Chips*, you took 15 samples of size 10 from each of three different populations, and graphed an experimental sampling distribution for each.
   a) What is the mean μ and standard deviation σ for each population? (You were asked to compute these in the Read and Study).
   b) Compute the theoretical means and standard deviations for the sampling distribution for samples of size 10 for each of these populations, according to the Central Limit Theorem.
   c) Now compute the mean and standard deviation of each of your 3 sampling distributions that you obtained in the class activity. How do your experimental results compare to the theoretical results of the Central Limit Theorem?
   d) Which population had the largest variation in sample means? Which had the smallest? Explain how the Central Limit Theorem would predict this result.

2. In *A Dicey Situation*, you took 30 samples of various sizes from a single theoretical population of the roll of a die.
   a) What is the mean μ and standard deviation σ for the population of all theoretical rolls of a single die?
   b) Compute the theoretical means and standard deviations for the sampling distribution for samples of size 2, 5, and 10 for this population, according to the Central Limit Theorem.
   c) Now compute the means and standard deviations of each of your 3 sampling distributions that you obtained in the class activity. How do your experimental results compare to the theoretical results of the Central Limit Theorem?
   d) Which sample size had the largest variation in sample means? Which had the smallest? Explain how the Central Limit Theorem would predict this result.

3. A population of kids has an average attention span of 15 minutes. Which is more likely, that one child chosen at random has an attention span of over 30 minutes, or that a group of 6 kids has an average attention span of over 30 minutes? Explain.

4. In figure 1, we drew a histogram representing the theoretical distribution of the mean of two independent rolls of a fair six-sided die.
   a) Calculate the theoretical probabilities involved; in other words, demonstrate the calculations that must have been used to create this histogram.
   b) This distribution can be viewed as the theoretical sampling distribution of the mean for samples of size 2 taken from the population of all rolls of a fair six-sided die. Use the probabilities you calculated in part a to find the mean and standard deviation of this theoretical sampling distribution.
   c) Now use the Central Limit Theorem to predict the mean and standard deviation of this theoretical sampling distribution. Do your results agree with part b? If not, can you explain the discrepancy?

5. Refer to figures 1-3 in the Read and Study.
   a) The histograms grew progressively taller from *Figures 1* to *3*. Why?

b) You have some freedom in choosing how to group the outcomes in a histogram. For instance, in place of *Figure 3*, we could have produced the histogram below. What are reasonable choices for the height of this bar, and why? Describe a disadvantage of this choice of a histogram.

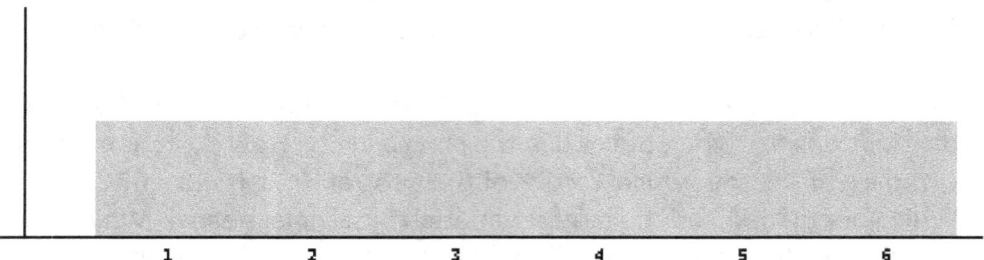

Chapter Three

Inference

Class Activity 14: Which Bag is Which?

There are three bags, labeled A, B and C on the bottom of the bag, so that when the bag is sitting on a table one cannot see the label. Bag A contains 10 white marbles, bag B contains exactly 5 red and 5 white marbles, and bag C contains exactly 10 red marbles. The instructor should then mix up the bags so that no-one knows which bag is which. One student should take a marble from just one of the three bags (without looking into the bag or looking at the label on the bottom. Based on this sample of one marble, the student should then guess whether they had selected from bag A, B, or C. The class should then discuss whether the student made a good guess and use probabilities to support their claims.

Now there is one bag, labeled X. *Nobody except the instructor should know the composition of this bag.* Each group should make a guess as to the proportion of marbles in this bag that are red. Then one student should take a sample of 10 marbles, with replacement, from this bag. Based on the results, discuss whether your group made a good guess. Use probabilities to support your claims.

Read and Study: Introduction to Inference

When you have eliminated the impossible, whatever remains, however improbable, must be the truth.

Sir Arthur Conan Doyle's Sherlock Holmes, in The Sign of Four

In *Section 1* we described *statistics* as the study of data collected from a *sample* in order to make inferences about a larger *population* from which that sample was drawn. In these final sections, we will explore methods of inference. The methods are nontrivial in nature, and are sometimes controversial, because they involve making assumptions about the population in order to draw conclusions about that same population!

In the *Class Activity* we drew a marble, or a sample of marbles, out of a bag in an attempt to make inferences about the contents of the bag. Here, the bags are representing a population about which we want to know a certain parameter, which we could call the proportion of the marbles in the bag that are red. Just like in doing real statistics, we don't know the true value of the parameter for the population we are sampling, and use the result of our sample to guess at what the population parameter might be.

In the first scenario, we had (unrealistically) some information about the possible distribution in each of the bags, and this allowed us to make some very precise probability calculations to support our guess. In fact, we could calculate the probability that our guess about the parameter was correct. In other words, *given our sample result, we could calculate the probability that our guess was correct.*

However, in the second scenario, we didn't know anything about the distribution of the population in the bag. This meant that we could not calculate the probability that our guess was correct. We could only content ourselves with calculating some hypothetical probabilities. If our guess was correct, we could figure out the probability that our sample result would occur. In other words, *given our guess was correct, we could calculate the probability of our sample results.*

This second scenario is the one that we find ourselves in most often when doing statistics. We don't know anything about our population, we only know what we got in our sample. Our approach here was to first make a hypothesis about what the true parameter was, then take a sample and use that sample result to assess the validity of that guess.

In the real world, we rarely know how many bags there are out there, or what they contain; we just see the metaphorical red marble that we happened to draw. We have to find ways to deal with this reality. We will investigate two good ways in the next two sections.

Homework

1) An instructor places two bags on a table – bag A containing 1 white and 2 red marbles, and bag B containing 1 red and 2 white marbles. *Nobody except the instructor knows which bag is which!* A student takes a marble from just one of the two bags (without looking into the bag), and gets a red marble. What is the probability that she picked from bag A?

2) Suppose you have 1000 bags, one (bag A) containing 10 red marbles, and each of the others containing 1 red and 9 white marbles. We randomly select a bag and draw a marble from it. Then the probability that we draw a red marble is small, only 10.09% (*Prove it!*). Suppose that we do, in fact, get a red marble. What is the probability that it came from bag A?

Side note: It is possible that our universe is only one of many, and that different laws of physics govern some of these universes. Some of the laws of our universe, such as those describing the nature of gravity, are required for our existence. So the fact that people are here to ask the question in the first place means we have to live in *this* type of a universe. (We have drawn a red marble.) This notion is known as the *anthropic principle* (ref 1). But whether ours is a high-probability type of universe or not, we do not know.

3) A Washington Post article (ref 6) states that among all births in China in the year 2000, the reported ratio of boys to girls was 117 to 100. This figure was not always so high; the "normal" ratio in China, as well as internationally, is approximately 105 to 100 (ref 6,7). There is plenty of evidence that something unusual is currently happening to the baby girls in China. The Post article argues that since the 1980s, strict governmental limits on births, combined with economic pressures and cultural values, have resulted in sex-selective abortions, as well as the medical neglect of and the hiding or underreporting of baby girls. (Traditionally in China, when a daughter marries, she becomes part of her husband's family and will no longer be there to take care of her parents. Thus, the parents may want to ensure that they have a son.) A U.S. Census Bureau report on a 1994 international symposium (ref 7) includes infanticide as another likely explanation for the absence of baby girls.

   a) China's year-2000 birth ratio of 117 boys to 100 girls could possibly be a fluke, due to random fluctuation. How could you begin to evaluate whether this is likely to be the case? What other information could help you to decide?

   b) In contrast to the explanations offered (sex-selective abortions, infanticide, etc...), suppose every family decided to have children until they had a boy, and then stop. Under this procedure, about half of the families would end up with one boy and no girls, so why wouldn't this also result in a high boy-to-girl ratio?

4) The following problem is included in an enjoyable book on probability (ref 9): "There are five apartments. One has two men, one has a woman and two men, one has two women and three men, one has six women and one man, and one has a married couple. If I knock on the door and a woman answers, what is the probability that I've reached the one with the married couple?"

We are to assume that the married couple is a man and a woman. The book's author presents the following solution: "This problem is deceptively simple. Since there are ten women who live in the apartment complex, and one of them is married, there is a 1/10 chance the woman you meet is the married one. So, the answer is 1/10."

a) The author's solution is incorrect. Explain what is wrong with it.

b) The book you are reading right now undoubtedly contains factual errors. Find some.

Class Activity 15a: A Loaded Question

The authors of this text work part-time as consultants for the Las Vegas Gaming Commission. They monitor the dice that casinos to make sure they are not using loaded dice to help tilt games in their favor. We have been sent six dice to investigate. Each group will be given one die. Investigate whether your die is fair. What do you suggest we report to the Gaming Commission? What evidence do you have that your die is fair or unfair?

Class Activity 15b: ESP Testing

Some people believe they have "ESP" (extrasensory perception), an ability to perceive communications by means of something other than the known senses. Do you think you have it?

Everyone should pair off and conduct an ESP test on each other. The test will consist of 30 questions, so the test-givers and the test-takers will need to prepare by numbering their papers 1-30. While the test-taker is not watching, the test-giver should create an "answer key," the answer to each question being one of the three letters A, B, or C, randomly chosen each time by rolling a die.

Now the test should begin. Place a barrier between each test-giver and test-taker, so that the test-taker cannot see the answer key. For each question, the test giver should concentrate on the appropriate letter until the test-taker has perceived the answer and written it down. When you have completed the 30 questions, switch places so that everyone gets tested. After the tests have been completed, they should be graded, and the results should be recorded for the class to see.

Does anyone in the class have ESP? How do you know?

Read and Study: Hypothesis Testing

You have learned that data from a *sample* can be summarized with a single representative value -- for instance, a *sample mean* or a *sample proportion*. Such a value is called a **statistic**. We really desire the corresponding value that represents the entire *population* -- this value is called a **parameter** -- but we don't usually have access to that much information. Examples of a population parameter include a *population mean* or a *population proportion*. (For the ESP test in your class activity, you might think of your personal sample statistic as the proportion of questions that you got correct, and your population parameter as the proportion of questions that your natural tendency would allow you to get correct in the long run.)

On a summer afternoon in the 1920s, scientist R. A. Fisher developed a method of deciding, based on data, whether a claim is likely to be true. He was attending a social gathering at which one of the ladies claimed she could distinguish whether first milk and then tea were poured into a cup, or first tea and then milk. Fisher put her to the test, repeatedly serving her tea with milk, and asking her to identify in which order they were poured. (A more thorough account of this summer afternoon, including a mention of how the lady fared, can be found in David Salsburg's book *The Lady Tasting Tea* (ref 1).) Fisher was developing a method now known as **hypothesis testing**. The basic procedure is this:

First you make some hypothesis regarding a population parameter. This is usually based on an assumption of neutrality, and is called the *null hypothesis*. (In the tea and milk experiment, the null hypothesis might be that the lady could not tell the difference -- that the true proportion of cups for which she would guess correctly is 1/2.) Then you gather data and compute the sample statistic. If the result is extremely contrary to what was predicted by the null hypothesis, then you reject the null hypothesis as incorrect.

How do you determine whether your sample result is "extremely contrary" to the null hypothesis? You pretend that the null hypothesis is true. Then you evaluate how likely it would be for a sample to result in a statistic as far away as yours was from the hypothesized value. This probability is called the **p-value** for your test. If, in your opinion, that the p-value is very low, then you are saying, "The result that I observed probably should not have occurred, if the null hypothesis were correct. But it did occur. Therefore, I think the null hypothesis is incorrect."

Note: Different people will have different opinions about what is a "very low" likelihood, so the procedure is highly subjective. The most common standard is that anything below a 5% chance is "very low." The standard that you are using for how low is considered very low is called the **level of significance** for the test. Sample results are called "statistically significant" if the p-value is below this level of significance.

Note: There are actually two hypotheses involved in a hypothesis test: the null hypothesis, which we have described; and the *alternative hypothesis*. We have simply taken the alternative hypothesis to be "the null hypothesis is incorrect." There are other choices, but we will not worry about them in this course.

Let's illustrate the method with our introductory example. Suppose the lady was given 20 different cups of tea and milk, some with the tea poured first and some with the milk poured first. As stated, we'll take the null hypothesis to be that the true proportion of cups for which she would guess correctly is 1/2. Under this assumption, the theoretical distribution for the sample proportion is represented by the histogram in Figure 1. This is a binomial distribution with 20 trials and a probability of success of 1/2, the same as the distribution of the proportion of heads in 20 tosses of a fair coin.

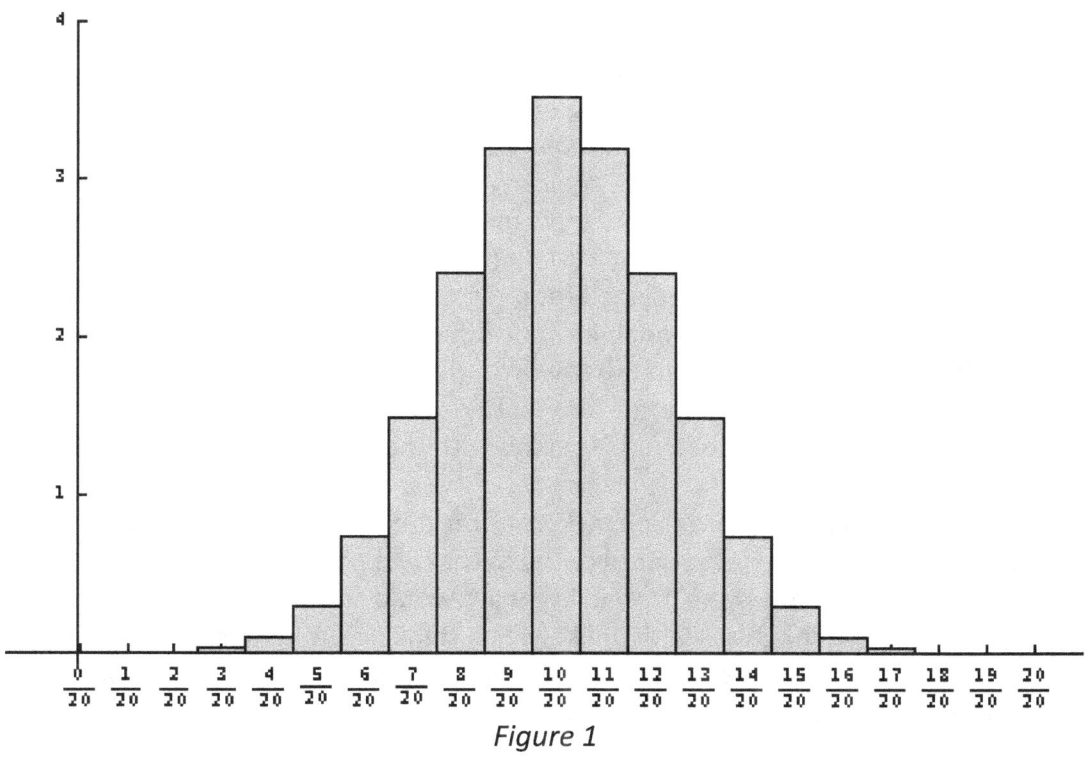

Figure 1

According to the null hypothesis, the true proportion is ½. In other words, in a sample of size 20 guesses we would expect half, or 10 out of 20 of her guesses to be correct. Notice how this expected value is also the mean of the sampling distribution shown above. The probabilities of the various other possible proportions of correct guesses, for theoretical samples of size 20, from left to right across the bars, are:

.0000, .0000, .0002, .0011, .0046, .0148, .0370, .0739, .1201, .1602, .1762, .1602, .1201, .0739, .0370, .0148, .0046, .0011, .0002, .0000, and .0000 (rounded to the nearest ten-thousandth).

Under the standard that any probability less than 5% is very low, you would reject the null hypothesis if the lady guessed correctly fewer than 6 times or more than 14 times out of 20, because the probabilities under those 12 bars add up to less than 5%. Such extreme results indicate that her ability differs *significantly* (either for the worse or for the better) from that of a 50% guesser. In these cases, statisticians would say that the sample result was statistically significant. But if she guessed correctly between 6 times and 14 times out of 20, then her result would be among the 95% of predicted sample results closest to what we would most expect (namely 10/20 correct), so you would not be persuaded to reject the null hypothesis; there is too good of a chance that she could have achieved such a result by random luck. In these cases, we could say that her results were not statistically significantly different from that of a 50% guesser.

The logic of hypothesis testing takes a little bit of getting used to, so we will use some analogies with other situations where you would use a probability to decide whether or not to reject a hypothesis. Suppose you work in a big office with many coworkers, and this office does not have any windows. When you came in to work in the morning, the weather was sunny, so you work with the assumption that it is not raining outside. Let's consider that to be our null hypothesis: that it is not raining outside. Suppose now you see a coworker enter the office holding an umbrella. This is some evidence against your null hypothesis. However, you might not consider this very strong evidence. Your logic might be as follows: "If it is not raining, what is the probability that I would see someone carrying an umbrella? Perhaps fairly low. But maybe it's just threatening rain, or rain is forecasted for later, and that's why that person has an umbrella. So I think there is still a reasonable chance that someone would by carrying an umbrella even if it were not raining." In other words, the sample result of seeing someone with an umbrella is not strong enough evidence for you to reject the null hypothesis that it is not raining outside. You decided that seeing the person with the umbrella was not significant.

But now what if instead you see three co-workers enter the office and they are all dripping wet. Now you might be inclined to reject your hypothesis that it is not raining outside. Now the logic is this: "If it is not raining, what is the probability that I would see three co-workers enter the office all dripping wet? I think this is very unlikely. So I think now that it is raining outside." In other words, if the null hypothesis were true, the sample results you saw would be very unlikely, and this causes you to reject your null hypothesis. You decided that seeing the three people dripping wet was significant.

Let's discuss one more example, but this time another real-world example of statistics in action. In section 2, we gave you data on a study of whether AZT had an effect on whether HIV was passed on from the mother to their child. In that study, 13/180, or about 7% of the AZT group passed on HIV, while 40/183, or about 22% of the control group passed on HIV. So there is an apparent relationship between getting the drug AZT and having a reduced HIV transmission rate. In section 2, we discussed 4 possible explanations for an apparent relationship between two variables: sample bias, confounding variables, random chance, and there really is a relationship. In a clinical study such as this, where participants were randomly assigned to the AZT and control groups, it is very unlikely that there are any sample bias or confounding variables influencing the results. One remaining concern is random chance. In order to say that we really think that AZT has an effect on HIV transmission, we need to consider this question: could it be that AZT really has no effect on HIV, and the study results are just due to random chance? In other words, could the AZT group and the control group really be just two samples from the exact same population, and by random chance one group got a low proportion of HIV transmission and the other group got a high proportion of HIV transmission?

To help us decide, we can conduct a hypothesis test. If the drug AZT truly has no effect on HIV transmission, then from the pooled data in the study we might estimate that the true proportion of infants that get HIV from their mothers might be 53/363, or about 15%. Using this as our null hypothesis, we can look at the sampling distribution samples of size 180 from this population. This is a binomial distribution with 180 trials and probability of "success" 0.15, shown in the figure below.

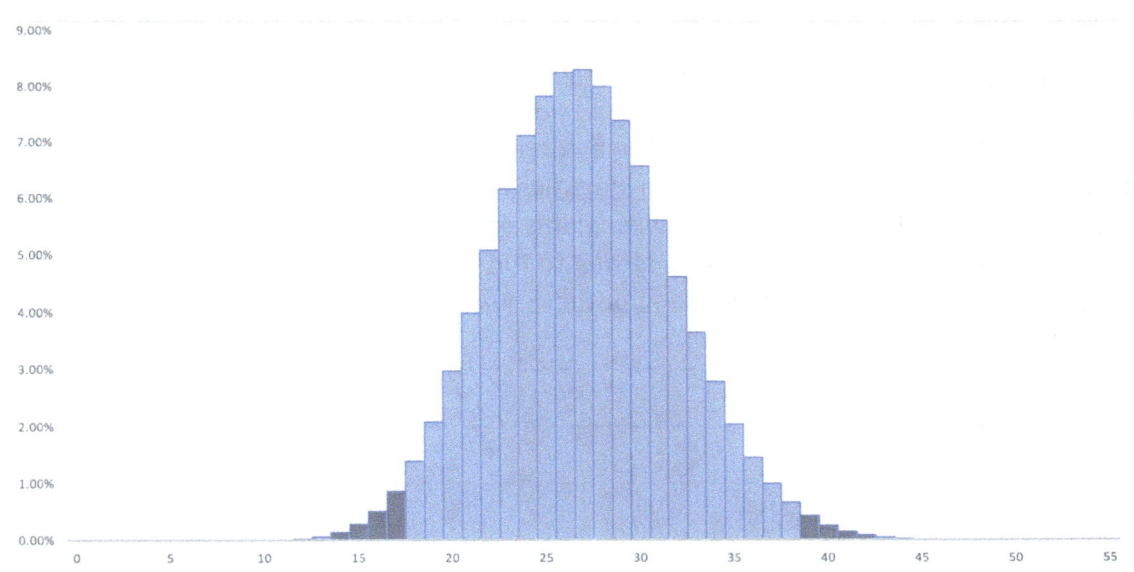

Number of HIV transmissions in a sample of size 180, if p=0.15

This distribution represents what we would expect to happen we took a random sample of 180 mothers with HIV and the probability of having a HIV transmission to the infant was really 0.15. We'd expect that the most common sample results would be close to the expected value, which is about 27 infants with HIV. But we can also see that it would be also relatively common to get samples with, say, 32 infants with HIV, but very unusual to get a sample with as high as 45 infants with HIV. By adding up the probabilities of the bars, we find that the interval from 18 infants up to 38 infants represents the middle 96% of samples. These bars have the lighter shading in the graph above. Any sample result that is fewer than 18 or greater than 38 infants would be in the 4% of samples that are the farthest away from the expected proportion. In other words, a sample result in this darker shaded region would have a p-value below 4%.

In the actual study, the AZT group had only 13 infants get HIV out of the 180; these sample results lie among the 1% of most extreme sample results, and so are statistically significant at a level of significance of 1% (the p-value is less than 1%). Thus the researchers in this study were able to conclude that the HIV transmission rate in the AZT group was statistically significantly different than in the control group. Such a low HIV transmission rate in the AZT group is not likely at all to happen just by random chance, making the result of this study very powerful evidence of the effectiveness of the drug AZT in reducing HIV transmission rates.

You could think about the hypothesis testing procedure as being similar to our criminal court trial procedure. Juries are instructed to judge a defendant guilty only if they believe there to be proof beyond reasonable doubt. To make the analogy complete, in a criminal trial, the null hypothesis is that the defendant is innocent. The "reasonable doubt" standard is the level of significance. Deciding to convict since they are guilty beyond a reasonable doubt is like deciding to reject the null hypothesis since the *p*-value is below the level of significance.

The criminal trial system is not designed to ascertain whether the defendant is more likely to be innocent or guilty; it is designed so that we should not convict innocent people very often. There is an assumption of innocence unless the evidence is quite strong to the contrary. Likewise, our hypothesis testing procedure in statistics is not designed to ascertain whether the null hypothesis is more likely to be right or wrong; it is designed so that this hypothesis should not be rejected unless the evidence is quite strong to the contrary. Hence, a failure to reject the null hypothesis does not imply that the hypothesis was correct -- it just says there wasn't enough evidence against it!

Homework

1) A certain midwestern university has approximately 12,000 students, 60% of which are female and 40% are male. Suppose that the Chancellor's office recently sponsored a contest where 6 randomly selected students are given free tuition. When they announce the results of the drawing, 5 of the 6 winners are male. Based on these results, some students complain that the drawing procedure was not truly random. What would you say? Support your answer by computing the probability that a random sample of 6 students would contain at least 5 male students. Calculate this probability two ways:

   a) Assume a student population of exactly 12,000, and use a sample space of all possible samples of 6 students.

b) Think about selecting 6 students successively from the student population. Assume the population is large enough so that every time a student is selected, the probability that the student is male remains 40%.

2) Suppose you give a 10 question multiple-choice exam, and each question has 4 possible choices. How many questions would someone need to get right before you would say they did statistically significantly better than just random-guessing? Use a 5% level of significance.

3) Pip, an incorrigible gamer, has manufactured a six-sided die that looks just like a standard one, but rolls 6 more often than a fair die. (He feels that this gives him an advantage at children's board games.) In order to escape detection, he has carefully balanced the die so that the theoretical mean of its outcomes is 3.5, the same as for a fair die.

   a) Explain how he could have achieved this; that is, assign probabilities to the outcomes 1 through 6 so that the probability of 6 is greater than 1/6, but the mean of the distribution is still 3.5.

   b) We, the authors, have just simulated 60 rolls of a die. The results are:

   > 1 appeared 14 times;
   > 2 appeared 11 times;
   > 3 appeared 6 times;
   > 4 appeared 7 times;
   > 5 appeared 8 times;
   > 6 appeared 14 times.

   We were using one of Pip's dice, but one that he claims to be a fair die. Perform an appropriate hypothesis test to see if you have enough evidence against Pip's claim. Carefully describe your null hypothesis and your test procedure. Use a 5% level of significance. What is your conclusion?

4) Here is a completely made up exercise: The National Association of Intellectual Vegetarians (this is not a real organization, as far as we know) asserts that 10% of Americans are vegetarians. You conduct a survey with a random sample of 60 Americans and find that only 3 are vegetarian. Does this data indicate that the proportion of Americans that are vegetarian is significantly less than 10%? Conduct a hypothesis test to support your answer.

5) Some people regard Ted Williams to be the best hitter in the history of baseball. His lifetime batting average was 0.344 for regular-season play over a Major League career spanning from 1939 to 1960. ("Batting average" is defined as the number of hits divided by the number of at-bats. So it is more accurate to think of this as a batting proportion: the proportion of at-bats that result in a hit.) His only World Series experience was in 1946; he got 5 hits in 25 at-bats, yielding a World Series batting average of just .200 (ref 2). Conduct

a hypothesis test to evaluate how strong this evidence is that his hitting abilities were not the same in the World Series as in during his regular season career.

6) After the 2002 season, the NFL considered whether they should change their overtime rules. The current rules at that time were that in overtime, the first team to score wins the game, and which team gets the ball first is determined by a coin toss. Opponents of this system claimed that whichever team wins the coin toss and gets the ball first has the advantage, since if they score on this first possession, they win without the other team getting a chance to have the ball. Their evidence was the data from the most recent season (2002) in which 16 times the winner of the overtime coin toss won the game, and only 9 times the loser of the overtime coin toss won the game. Does this data indicate that winning the coin toss has a statistically significant effect on winning the game? Support your answer by conducting an appropriate hypothesis test.

Read and Study: The Normal Distribution

Recall that the Central Limit Theorem says that as samples grow larger, the distribution of the sample means becomes bell shaped. Since a proportion is an example of a mean (you have established this through some old homework exercises), this result applies as well to sample proportions. *Figure 2* below reproduces the histogram from *Figure 1*, for the binomial distribution of the proportions of teacups correctly identified, and superimposes the normal distribution which approximates it.

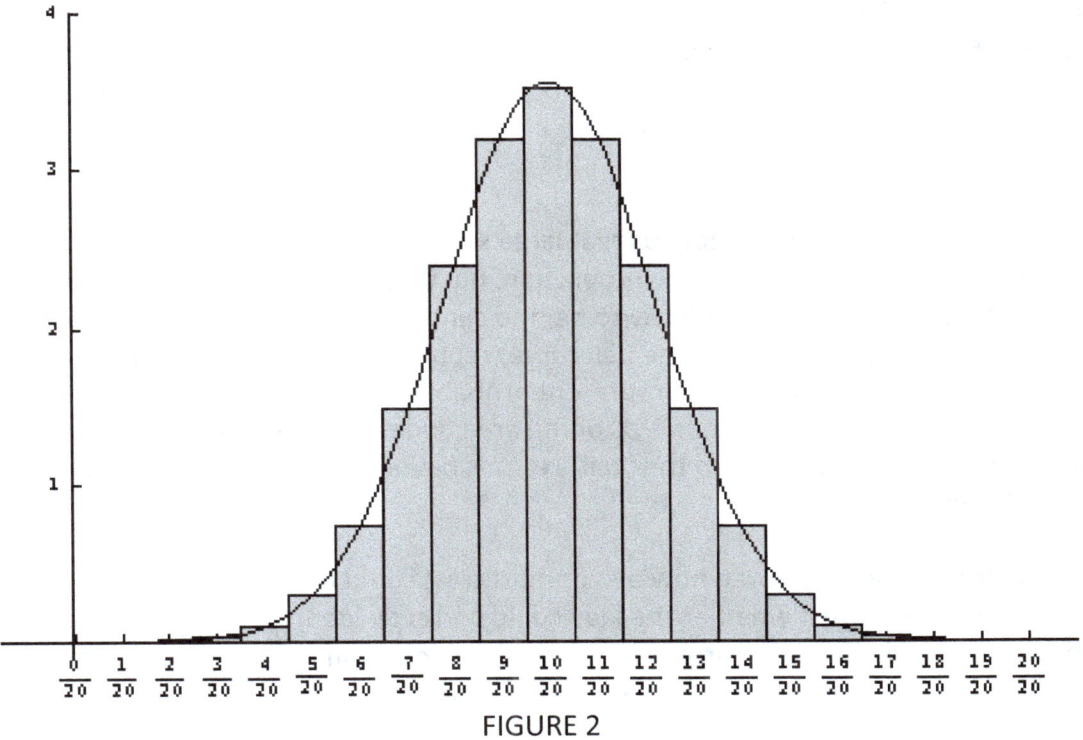

FIGURE 2

In general, suppose *p* represents a true (population or theoretical) proportion and *n* represents the sample size of the samples in the sampling distribution. Then for large sample size, the binomial distribution will be approximately a normal distribution with a theoretical mean *p* and theoretical standard deviation

$$\sqrt{\frac{p(1-p)}{n}}.$$

Note: A history of the development of the normal distribution is presented in (ref 5).

Problem 1: Consider flipping a fair coin once, and consider the distribution of the number of heads. This is a binomial distribution with a sample size of 1. Compute the mean and standard deviation of this distribution. Do not use the formula given above, instead, compute it from scratch.

Problem 2: Consider tossing a fair die once, and consider the distribution of the number of *ones*. This is also binomial distribution with a sample size of 1. Compute the mean and standard deviation of this distribution. Do not use the formula given above, instead, compute it from scratch.

Problem 3: Consider one trial where the probability of success is p, and consider the distribution of the number of *successes*. This is also binomial distribution with a sample size of 1. Compute the mean and standard deviation of this distribution. Do not use the formula given above, instead, compute it from scratch.

The law of large numbers already told us that large samples will tend to result in sample proportions which are close to the true proportion, but it did not give us a way of measuring how close the sample proportion is likely to be; the central limit theorem does, because areas under portions of the normal curve (the bell curve) represent probabilities, and those areas can be measured. (Recall that we likewise represent probabilities by areas, rather than heights, in histograms.) It so happens that about 95% of the area under a normal curve lies within 2 standard deviations of the mean; so the "tail" regions beyond 2 standard deviations account for only about 5% of the total area.

Problem 4: Previously we showed how a hypothesis test could be carried out using the binomial distribution to test whether the lady could correctly identify teacups. If we instead used the *normal* (bell-shaped) approximation to this distribution, under what circumstances would we reject the null hypothesis (again using the 5% standard)?

The rules of gene inheritance theory were introduced in the 1800's by Gregor Mendel. He hypothesized that the flower color in pea plants results from a pair of genes (one inherited randomly from among the two genes of one parent, and the other inherited randomly from among the two genes of the other parent), and that the genes for purple flowers were dominant to those for white flowers. Thus, a white-flowered child must have inherited two white genes, but a purple-flowered child could have inherited either two purple genes or one white and one purple.

Problem 5: In preparation for a test of his theory, Mendel created a generation of purple flowers, each having one white and one purple gene. How did he do this? (How could he know whether a particular purple flower had genes of each color, as opposed to two purple genes?)

Problem 6: He then bred this generation of mixed-gene purple-flowered plants with each other. This resulted in 929 offspring, 224 of which ended up with white flowers, and 705 with purple flowers. Using the null hypothesis that supports his theory, carry out a hypothesis test on Mendel's data.

Note: (The information in this problem can be found in ref 6 and 7.) Time and time again, Mendel's experimental results provided strong support for his theory. In a study published in 1936, R. A. Fisher wondered how likely it was that Mendel's data should have agreed so closely with his desired values. He computed the probability of such good agreement to be about 0.00007, and concluded that Mendel's data are probably faked. Why would Mendel have faked his data? Possibly because until the development of modern statistical methods, people did not have a way of interpreting how close an experimental result should be to the theoretical result; nobody knew what was an acceptable margin of error. Perhaps Mendel believed that any experiments which did not give strong confirmation to his theory must be flawed, and he may have found excuses to throw away some of his data. Other notables who probably faked their data include Isaac Newton and Galileo Galilei. (See ref 8.)

Class Activity 16: Yanny or Laurel?

Your instructor is going to play an audio clip for you to listen to, of someone repeatedly saying a word. (Instructors, you'll easily find the audio clip with an internet search for "yanny or laurel".)

Did you hear the word "yanny" or "laurel"? Take a vote in your class to see how many people heard which word.

Let's assume now that our class constitutes a random sample of all pre-service middle school math teachers at our university (the population). Work with your group to find a range of plausible values for the true proportion among the population that would hear the word "yanny".

Note: Of course the range 0% to 100% works, but that doesn't tell us anything. You'll want to keep the range as narrow as possible (you're trying to make an accurate prediction), but wide enough that you think it would probably contain the true proportion.

Here are some additional things to think about:
- What are some factors that could affect the reliability of our study?
- What if our population of interest was all pre-service middle school math teachers enrolled in American universities, instead of just at our university? World-wide universities?

Read and Study: Confidence Intervals

In *Section 15*, we explored one method of making inferences about a population, upon observing a sample. We did this by starting with some preconceived hypothesis about the *population parameter*, and checking whether our observed *sample statistic* was compatible with that hypothesis. In this section, we are exploring another method of making inferences about a population. This time, rather than beginning with a preconceived hypothesis, we will treat our sample statistic as our best estimate for the population parameter; then we will evaluate how close that estimate is likely to be.

A CNN / *USA Today* / Gallup poll conducted from March 10-12, 2006 asked 1,001 American adults whether they approved of the way George W. Bush was handling his job as president (see ref 1). 36% of those polled said they approved. CNN reported,

> "One can say with 95% confidence that the margin of sampling error is ± 3 percentage points."

This makes it sound as though we can be 95% confident that the *true* proportion (among the population of all American adults) is somewhere between 33% (that's 36% - 3%) and 39% (that's 36% + 3%). But the wording can be misleading. A more honest interpretation of the poll is this:

> The sample statistic they got from their poll, namely 36%, is likely to occur among a sample of 1,001 people, *provided* the population parameter (i.e., the true proportion) is between 33% and 39%. Conversely, the observed 36% statistics would be unlikely to occur if the true figure were outside of the 33%-to-39% range.

The "95% confidence" really refers to the cutoff between what we consider *likely to occur* and *unlikely to occur*.

Let's analyze this scenario further so that we can truly understand that last paragraph. Suppose the *true* approval proportion were really 33%. We know that when sampling from a population that the sample results we observe will vary from sample to sample. The histogram in *Figure 1* shows the distribution of possible sample values for samples of size 1,001. (This is a binomial distribution with 1001 trials and a probability of "success" of 0.33.)

You can see that for such a large number of trials, the distribution of these samples looks very normal. Naturally, a sample result of 33% is most common, but 34% is also very common, and so is 35%. However, it would be extremely rare to get a sample result as high as 38%. To make a distinction between which sample results are "likely" to occur and which are unlikely, we will say a sample result is likely if it is one of the 95% of all samples most closely surrounding the true parameter. So in figure 1 we have shaded this 95% of the distribution.

Notice that the approval proportion from the CNN / *USA Today* / Gallup poll, 36%, lies right at the edge of this region. If the true proportion were less than 33%, the distribution would shift farther left, leaving the sample result of 36% outside the shaded region. In other words, if the true proportion were less than 33%, CNN's sample result would be considered very unlikely to happen.

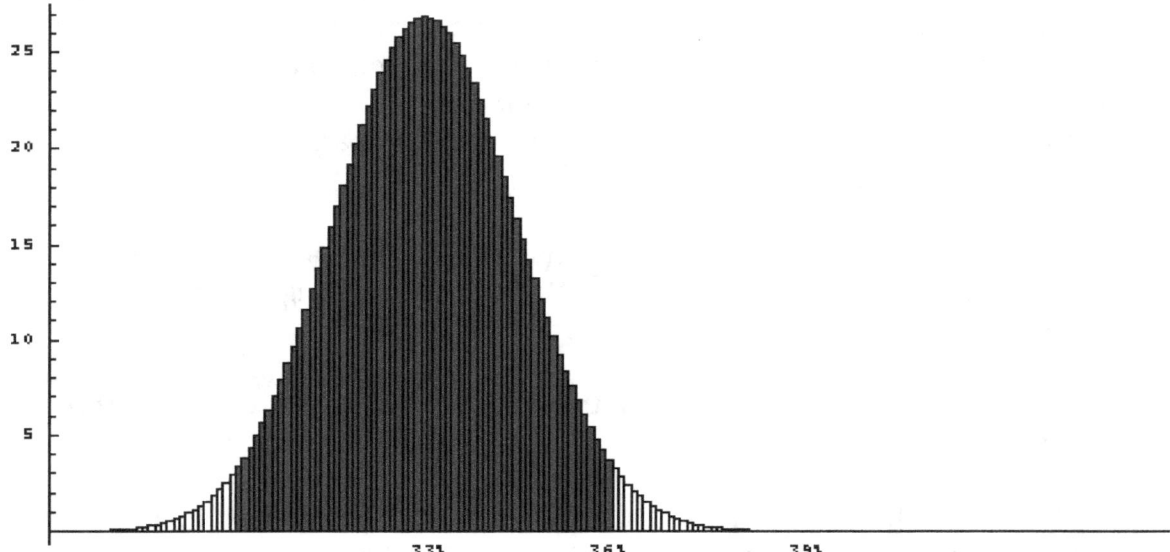

Figure 1 (assuming a *true* proportion of 33%)

The other extreme, supposing the true approval proportion to be 39%, is similarly depicted in *Figure 2*. The sample result of 36% also lies within its shaded region. But if the true proportion were anything higher than 39%, the distribution would move father right, leaving the sample result of 36% outside the shaded region. In other words, if the true proportion were greater than 39%, CNN's sample result would be considered very unlikely to happen.

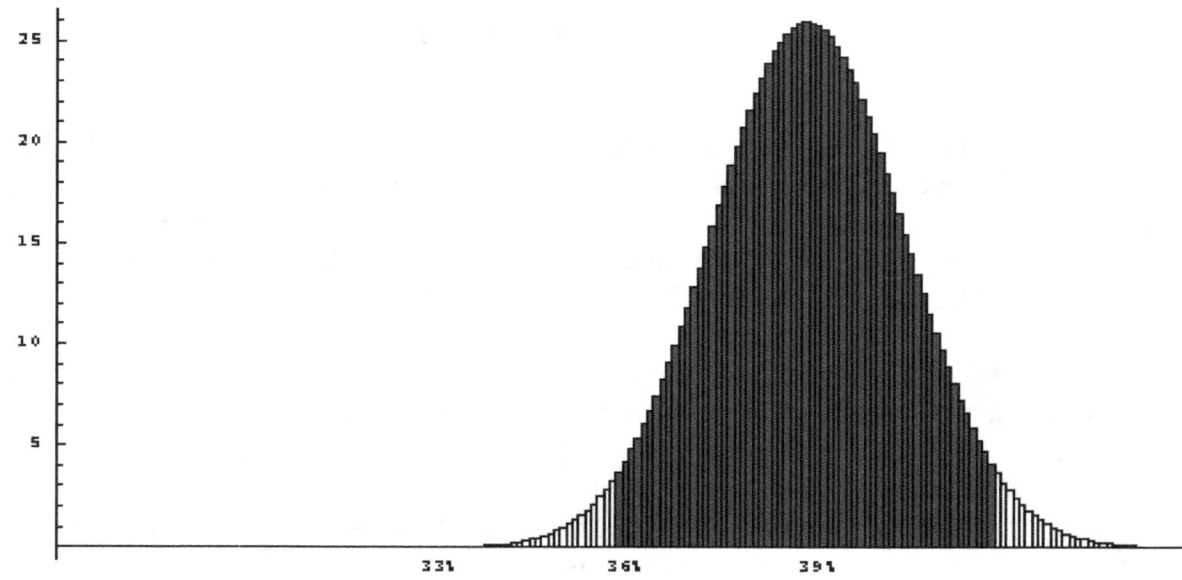

Figure 2 (assuming a *true* proportion of 39%)

The interval from 33% to 39% is called the *95% confidence interval* for the true approval proportion. Summarizing the general idea: You obtain a statistic from a sample. Then you find an interval of possibilities for the true parameter. For each possibility in this interval, the *observed* statistic must be among the 95% of possible sample statistics lying closest to that parameter. This interval is called a 95% **confidence interval** for the parameter. (The 95% level, incidentally, is entirely arbitrary but usually the standard.)

Here's another way of thinking about it, in the context of hypothesis testing: the confidence interval is the collection of values which could have been used for your null hypothesis, without a rejection based on your observed statistic. (A 95% confidence interval corresponds with a 5% rejection standard.)

The method of confidence intervals was developed in the 1930s by Jerzy Neyman (ref 2). We will see through some of our homework problems that "confidence" may not mean exactly what we'd like for it to mean; Neyman made it clear that the confidence level applies to the *process*, rather than to each result. Nevertheless, the concept is difficult enough that it is often misunderstood. In fact, some prominent statisticians strongly objected to the idea when it was developed, perhaps because they didn't understand it fully. (So it's normal for you to feel overwhelmed as you try to understand this material. You might have to reread this section a few times as you work on the homework problems. In fact, a lot of mathematical ideas are difficult like this. All mathematicians know first-hand how difficult and frustrating mathematics can be sometimes – for *all* of us.)

In addition to misrepresenting the nature of confidence intervals, some treatments make it seem as though the topic must be presented via the *normal distribution.* Assuming a normal distribution allows one to easily calculate a confidence interval using some simple formulas that bypass the logic of the confidence interval that we have tried to explain in this section. Our goal is to understand the meaning of a confidence interval, and so we will include problems involving proportions that may be solved through simple calculations using the *binomial distribution*. Identifying the confidence intervals in this direct way for small samples involves more judgment calls, because the distributions are often not symmetrical, and the cutoff values may not be sharp -- but the fundamental ideas are there, and a better overall understanding will be developed (we hope).

Homework

1. You randomly selected 5 Americans and asked them whether they approved of the way the president was handling his job, and that 2 of them said yes.

    a) Suppose that, in reality, 10% of all Americans approved of the way the president was handling his job. Make a histogram that shows the distribution of the various possible outcomes for a survey of 5 randomly-chosen people in this case. Verify that your observed sample proportion of 3 out of 5 would be among the 95% of possible resulting proportions lying closest to the true proportion (10%).

b) Suppose instead that 80% of all Americans approved of the way the president was handling his job. Draw a histogram depicting the distribution of the various possible outcomes for a survey of 5 randomly-selected Americans in this case. Then, as you did in *part a)*, verify that a resulting sample proportion of 3 out of 5 would be among the 95% of possible resulting proportions lying closest to the true proportion (80%).

Note: You have just worked with the low and the high ends of a confidence interval based on a survey of 5 people, out of whom 40% (2 people) approved. Upon completing this exercise, you will have shown that a 95% confidence interval around that figure would extend from about 10% to 80%.

2. Suppose you have obtained a statistic from a sample. Which is larger, the 95% or the 99% confidence interval around that statistic? Explain your reasoning. Would a confidence interval shrink or grow if you took a larger sample? Explain your reasoning.

For the next two problems a binomial distribution computer application will be needed to be able to quickly compute probabilities for various binomial distributions. The one at http://www.stat.berkeley.edu/~stark/Java/Html/BinHist.htm) is an especially nice one.

3. Suppose you take a random sample of 20 students at your university, and 14 of them own a smartphone. Thus the sample proportion is 0.70, or 70% of the sample owns a smartphone. Would 0.60 be in the 95% confidence interval for the true proportion of students at your university that own a smartphone? How about 0.90? Estimate the 95% confidence interval for the true proportion to the nearest tenth.

4. Consider taking a public opinion poll of 100 randomly selected likely voters. Suppose that 40 out of the 100 report that they will be voting for the Democratic candidate in the coming election, resulting in a sample proportion of 0.40. Verify that the 95% confidence interval for the true proportion of likely voters that will be voting for the Democratic candidate would be approximately from 0.30 to 0.50. *Verify this*. Thus the results of this poll could be reported as saying that 40% will be voting for the Democratic candidate, with a margin of error of ± 10 percentage points. Suppose now that the poll was of 500 likely voters, and that the sample proportion was again 0.40. Now what is the margin of error? What if the poll was of 1000 likely voters and the sample proportion was again 0.40. What is the margin of error then?

Class Activity 17: Fingerprints

Materials needed: Scotch tape, a pencil, and several sheets of paper for each group.

ARCH WHORL LOOP

A fingerprint can usually be classified as an *arch*, a *whorl*, or a *loop*. Each of these types is illustrated above. Since about 1900, law enforcement agencies have identified criminals through fingerprints based on an arch-whorl-loop analysis (ref 1). Modern procedures are more sophisticated than what we are about to do, but we can all discover our simple fingerprint signatures. On a blank sheet of paper, draw an outline of your right hand. With the pencil, darkly shade a 1-inch square region on a blank spot of paper. Rub your right thumb on this spot; then touch your thumb to the sticky side of a piece of tape, and stick that tape to the appropriate spot on the outline of your hand. You should clearly see your thumbprint. Do this for each of the fingers on your right hand. Using the illustrations at the top of this page as a guide, label each print as either an arch (A), a whorl (W), or a loop (L). Reading from left to right, beginning with your thumb, you now have a right-hand signature, such as "L-L-A-L-W".

The relative frequencies of fingerprint types is commonly accepted to be 5% arches, 30% whorls and 65% loops (ref 1 or 2). Suppose one of you committed a crime at your university, and the police found a print of the culprit's right hand at the scene, and classified it using the arch-whorl-loop system. Then they began fingerprinting the right hands of all students at the university, until they found someone whose prints matched, at which point they abandoned their search and arrested that person as the sole suspect.

If the police selected the students in a random order, what is the probability that the wrong person (ie, someone other than the culprit) was arrested? Does the answer depend on which one of you is the culprit?

If the police selected the students alphabetically from the student directory, what is the probability that the wrong person was arrested? What makes this question different in nature from the previous?

Assuming that the police made no mistakes in classifying the handprints, how could the fingerprinting procedure be so unreliable in identifying the correct suspect?

Read and Study: Bayes' Theorem

During the half-time show of a basketball game, one lucky fan named Hoop is given the opportunity to win a million dollars. He begins by tossing a fair coin. If it lands *heads*, he takes a shot from the free throw line (where he happens to be a 75% shooter); if it lands *tails*, he takes a shot from half court (where he happens to be a 10% shooter). If he makes the basket from whichever position he was assigned, he wins a million dollars. Suppose Hoop ends up winning the million dollars. *What is the probability that he took the half-court shot? Think about this for a few minutes. See if you can figure it out.*

We know the probability of Hoop's making the shot, given that he took it from half court, to be 10%, but we want to know the "inverse" probability. A method for solving such problems was developed in the 18th century by the English reverend and amateur mathematician Thomas Bayes. **Bayes' Theorem** presents a formula for inverting conditional probabilities -- that is, for calculating *the probability of an event A, given that B occurred*, based on *the probability of B, given that A occurred* (together with some additional information). In his book *Calculated Risks* (ref 3), Gerd Gigerenzer makes a good case for presenting the theorem via frequencies rather than probabilities in introductory courses. To support his argument, he interviewed doctors, lawyers, and other professionals who had been exposed to the traditional form of Bayes' Theorem in their training. Almost none of them were able to remember what the theorem said, or to answer problems of the sort we posed in the preceding paragraph, although such problems were highly relevant to their professional work. However, they were much better at solving the same problems if the information was presented through frequencies rather than probabilities. We now proceed to solve our problem in this manner.

Suppose 200 people, each with the same shooting abilities as Hoop, were given the opportunity described. About 100 of them would toss *heads* and shoot from the free-throw line, and of these, about 75 would win the money. The other 100 would shoot from half court, and about 10 of them would win the money. Altogether, about 85 people would win the money, of which 10 would have taken the half-court shot. So the probability that a winner took the half-court shot is 10/85. (We'll see in *Homework # 1* that this probability could be very different if the coin were weighted so that Hoop shot the ball from half court more often. This is the sort of problem we were hinting at with the bag distributions in an earlier activity.)

In legal and medical contexts, laboratory tests are often run to indicate some condition -- the presence of a disease in a patient, for example, or a criminal suspect's match with the DNA from a crime scene. Sometimes the condition is present and the test fails to identify it; such a test result is called a **false negative**. Other times a test falsely indicates a condition that was *not* actually present; such a test result is called a **false positive**. A good test should have low rates for both false positives and false negatives.

Fingerprint analysis standards have not been established, and some high-profile cases have emerged where false fingerprint identifications led to arrests of innocent people (ref 4). DNA analysis standards have been established, but, contrary to public belief, test results are far from infallible. Reading an individual's DNA profile is a skilled art; the technician must decide whether certain things line up with each other. Answers could vary from one technician to another, or from one sample of an individual's DNA to another. Even for the best laboratories, false positive rates might be as high as 1 in 200 (ref 3). This might seem low, but imagine if all Americans were tested to see whether they matched a perpetrator's DNA from a particular crime scene. One out of every 200 Americans, or about 1.5 million people, might be identified as matches! Communities should not be genetically screened as a matter of procedure in criminal investigations; other means of identifying suspects should be used to narrow down the selection.

Testing for the presence of HIV (the virus which causes AIDS) provides a significant medical illustration. The Food and Drug Administration has approved certain home-testing kits. The allowable false-positive rate is 0.5%, or again, 1 in 200 (ref 5). At the end of 2003, slightly more than 1 million Americans were living with HIV/AIDS (ref 6). If all 300 million Americans took an HIV test, then in addition to approximately 1 million accurate positive test results, another 1.5 million people would obtain false positive results. So more than half of the results would be wrong, although the test was labeled "99.5% accurate." *Create a model or diagram explaining why this is the case. Explain how this situation is related to conditional probabilities.*

Homework

1) Refer to the opening problem from this section's *Read and Study*.

    a) Suppose the coin that Hoop tosses is weighted so that it lands *heads* only 1% of the time. Again, calculate the probability that he took the half-court shot, given that he won the money.

    b) Suppose 1,000 people are given this same opportunity, and that a fair coin is used. Suppose that all of the 1,000 people are 100% free-throw shooters, but that we don't know their skill levels from half court. If 600 people win the money, then approximately what percentage of the half-court shots were successful?

c) You are a government official and have been assigned to determine what percentage of people cheat on their income taxes. You will pose the question to a sample of 1,000 people -- but you are afraid that if you ask the question directly, none of them will admit to cheating, for fear of reprisal. Explain how you could use the idea from *part b)* to gather meaningful data in a way that would *ensure* anonymity.

2) The following information is summarized from (ref 7). On June 18, 1964, a Los Angeles woman was pushed to the ground in an alley and robbed of her purse; she described the perpetrator as a young woman with blond hair. A man who lived nearby heard screams, and then saw a Caucasian woman with a dark blond ponytail run out of the alley and escape in a yellow car being driven by a "Negro man" with a mustache and beard. Some other witnesses described the car as yellow with an off-white top. Four days afterwards, police questioned Malcolm Collins and his wife, Janet, regarding the robbery; they were later arrested and brought to trial. The defendants owned a yellow Lincoln automobile with a white top; Janet's hair was dark blond and she sometimes wore a ponytail; Malcolm sometimes wore a beard, though he did not have one at the time of questioning. In the course of their trial, the prosecutor used these identifying characteristics to persuade the jury of their guilt by means of a "mathematical proof." He based his argument on the following probabilities, which he presented merely as a matter of conjecture, and not corroborated by any statistical evidence:

Characteristic	Probability
Partly yellow automobile	1/10
Interracial couple in car	1/1000
Negro man with beard	1/10
Man with mustache	1/4
Girl with ponytail	1/10
Girl with blond hair	1/3

Multiplying these together, he got $\frac{1}{12,000,000}$, and said that this demonstrates a 1 in 12 million chance that the defendants were innocent.

a) The prosecutor's descriptions of the characteristics in the chart above are ambiguously written. Carefully restate the information on the chart, using conditional probabilities. What assumption did the prosecutor need to make in order to multiply these probabilities together? Do you think that assumption was valid?

b) Suppose the probability that a random couple would possess all of the above characteristics really was 1 in 12 million, after all. Does this then imply the prosecutor's conclusion of a 1 in 12 million chance that the defendants were innocent?

3) Random drug testing is a common occurrence in our society today. Athletes, job applicants and employees, and school children are often tested for illegal drug use. John P. Walters,

director of the Office of National Drug Control Policy, writes, "Testing has been shown to be extremely effective at reducing drug use in schools and businesses all over the country. As a deterrent, few methods work better or deliver clearer results." (ref 8) A fact evidently misunderstood by many policy makers is that the results are in reality *not at all* clear.

There are several different methods of testing for drug use. Two of the most common are EMIT ("enzyme multiplied immunoassay technique," which uses antibodies to detect the presence of drugs in the urine), and GC/MS (a "gas chromatography / mass spectroscopy" urine analysis). Though it is considered to be less accurate, EMIT is commonly used for initial screening because of its lower cost (ref 9).

Because of the possibility of false positives, it is generally recommended that people who test positive for drug use in an initial screening should receive a GC/MS follow-up test, but these recommendations are not always followed. A 1991 National Institute of Justice study (ref 10) reports that: among 732 people identified by GC/MS as users of cocaine, 167 falsely tested negative by EMIT; and among 1,916 people identified by GC/MS as nonusers of cocaine, 48 falsely tested positive by EMIT. (Note that the study adopts the conventional assumption that GC/MS is highly accurate. We, the authors, have not seen any statistical evidence supporting this assumption, however.) The U.S. Department of Health and Human Services estimated in 2003 that approximately 1% of Americans aged 12 or over were users of cocaine, within the past month (ref 11). There are roughly 250 million Americans aged 12 or over, as of the year 2005 (ref 12). Suppose all were tested for cocaine using EMIT. Based on the results of the National Institute of Justice report, approximately how many Americans would *falsely* test positive for cocaine under this procedure? How many would legitimately test positive?

4) The information cited in this problem is from (ref 13,14).

   a) A recent study involving 1,909 women compared the screening abilities of MRIs versus mammograms for detecting breast cancer. Among 45 cancers, 22 were identified by MRI but not mammography, 8 were identified by mammography but not MRI, and 10 were identified by both. How many were identified by neither? What percentage of the cancers were detected by MRI? By mammography? Is there any advantage to using *both* methods, as opposed to just one or the other?

   b) The same study reported a 10% false positive rate for MRI, versus a 5% false positive rate for mammography. Other studies have indicated that among women under age 50, mammograms detect only about 30% of breast cancers, but that among women over age 50, they detect 80-90%; the detection rates for MRI are not affected by age. An MRI costs about $1,300-$1,400, while mammograms cost about $110 (2004 figures). Approximately 275,000 women each year are diagnosed with breast cancer in the U.S. If you were in charge of public health policy, what recommendations would you make for women regarding breast cancer screening, and why? What *other* information would be helpful in guiding your decisions? Look up any relevant missing information and incorporate it into your analysis.

5) Red-green colorblindness is caused by a gene located on the X chromosome. Every girl inherits two X chromosomes, one from her mother and one from her father. She will be colorblind only if *both* of these chromosomes are defective. Every boy inherits one X chromosome (from his mother) and one Y chromosome (from his father). He will be colorblind if the single X chromosome is defective (because he has no second one to compensate). Approximately 1 out of every 12 X chromosomes is color-defective. (We are simplifying things somewhat; see ref 2, 3 for more information. You can also take a test for colorblindness at the website in ref 3.)

What fraction of males are colorblind? What fraction of females are colorblind? What fraction of colorblind children have a colorblind mother? How could this last answer be so low, since colorblindness is an inherited disorder? Justify your answers using probabilistic reasoning.

(You can simulate the situation: Get 2 paper bags, and 7 small plastic chips or paper slips. Label one of your paper bags "mother" and the other "father;" label one of your chips "Y," three "X-normal," and three "X-defective." Put the Y chip into the father bag; put one X-type chip into the father bag and two into the mother bag, each time choosing either X-defective with probability $\frac{1}{12}$ or X-normal with probability $\frac{11}{12}$. Randomly draw one chip from each bag to create a child. Experimenting with this may help you to understand the various possibilities.)

6) Genetic mutations sometimes result in inheritable diseases. If both genes in a person's relevant gene pair must be defective in order for the disease to be manifested, then the inheritance pattern is said to be recessive. A child inherits one gene as a random selection from his/her mother's gene pair, and the other as a random selection from his/her father's gene pair.

   a) Suppose that some environmental circumstance (such as a brief massive radiation exposure) caused such a genetic mutation, resulting in a recessive inheritance pattern disease, among a large group of people, so that a random selection of half of all the relevant genes in the pool were affected. Suppose also that this disease, when exhibited, was fatal. What fraction of the exposed people would die from the disease? Among those that lived, some would have only one mutated gene. (Such people are called carriers, because they could still pass the disease on to future generations.) Among the children of the survivors, what fraction would exhibit the disease?

   b) Recessive inheritance pattern diseases can affect many generations within a population, even when they are lethal. One such example is cystic fibrosis, a recessive-pattern inherited disease of the respiratory and gastrointestinal tracts; it commonly results in death before adulthood, so we will suppose that no children are conceived by those who exhibit the disease. Among children of European descent born currently, about 1 in 2,000 exhibits cystic fibrosis (ref 4). In example a), the fraction of people exhibiting the disease in the population dropped dramatically from one generation to the next. Is this also the case for cystic fibrosis? Explain.

c) Some diseases, such as Huntington disease (ref 5), follow a dominant inheritance pattern, meaning that only one gene in the pair need be defective in order for the disease to be exhibited. Some genetic mutations (not Huntington disease, however) cause sure fatality in newborns. Explain why no such dominant-inheritance-pattern disease that is lethal to infants has ever established itself among humans.

References

(Section 1: Feet and Hands)

1) Koestler, A., *The Watershed: A Biography of Johannes Kepler*, University Press of America, Lanham, Maryland, 1960.
2) Yahoo <http://finance.yahoo.com/>
3) Huff, D., *How to Lie with Statistics*, W. W. Norton & Company, New York, 1993 (originally copyrighted in 1954).
4) National Council of Teachers of Mathematics. (2000). *Principles and Standards for School Mathematics*. Reston, VA: Author
5) *Consumer Reports*, February 2005, May 2005, and September 2005.
6) Internal Revenue Service <www.irs.gov/pub/irs-soi/04db31ps.xls>
7) World Health Organization <www3.who.int/whosis/mort/table1.cfm>
8) College Board <http://www.collegeboard.com/prod_downloads/press/cost06/trends_college_pricing_06.pdf>
9) Economic Policy Institute <http://www.epi.org/publications/entry/tables_figures_data/>

(Section 2: HIV and AZT)

1) Connor, E. M., et al. "Reduction of Maternal-Infant Transmission of Human Immunodeficiency Virus Type 1 with Zidovudine Treatment." *The New England Journal of Medicine*, 1994; 331:1173-1180. <https://www.nejm.org/doi/full/10.1056/nejm199411033311801>

(Section 3: It's a Long Shot)

1) Hacking, I., *An Introduction to Probability and Inductive Logic*, Cambridge University Press, Cambridge, 2001.
2) Maistrov, L. E., *Probability Theory: A Historical Sketch*, Academic Press, Inc., New York, 1974
3) Bernstein, P. L., *Against the Gods: The Remarkable Story of Risk*, John Wiley and Sons, Inc., 1998
4) Metropolitan Transit Authority, New York City Transit <http://www.mta.nyc.ny.us/nyct/service/schemain.htm>
5) Robinson, S. "Why Mathematicians Now Care About Their Hat Color." *The New York Times*, April 10, 2001.

(Section 4: Spaghetti Triangles)

1) "Flipping, Spinning and Tilting Coins," Chance News 11.02, the Chance Team at Dartmouth College, 18 Feb. 2002 to 20 April, 2002, <http://www.dartmouth.edu/~chance/chance_news/recent_news/recent.html>
2) Feynman, R. P., R. B. Leighton and M. Sands, The Feynman Lectures on Physics, Volume I, Addison-Wesley, 1963.
3) <http://www.random.org/nform.html>
4) Shaughnessy, J.M. (1992). Researches in probability and statistics: Reflections and Directions. In Grouws, D.A. (Ed.). "Handbook of Research on Mathematics Teaching and Learning (pp.465-494)." New York: Michigan Publishing Company.
5) Paulos, J. A., A Mathematician Plays the Stock Market, Basic Books, New York, 2003
6) Hill, T., "The Difficulty of Faking Data," Chance 26, 8-13 (1999). Also available for download, under "Publications," on Ted Hill's website: <http://www.math.gatech.edu/~hill/>
7) Ulam, S. M., Adventures of a Mathematician, University of California Press, 1991

(Section 5: Hanging in the Balance & The Best Answer)

1) Cauchan, D. "Newspapers' Recount Shows Bush Prevailed." *USA Today*. May 15, 2001. <http://www.usatoday.com/news/washington/2001-04-03-floridamain.htm>
2) CNN.com, "In-Depth Special: Florida Ballots Project," 2001. <http://www.cnn.com/SPECIALS/2001/florida.ballots/stories/main.html>
3) Placket, R. L., "The Principle of the Arithmetic Mean." Biometrika **45** (1958), 130-135.
4) "Neonatology." *World Book Multimedia Encyclopedia*. World Book, Inc., 2001.
5) De Groot, M. H. and Schervish, M. J., *Probability and Statistics, 3rd ed.*, Addison-Wesley, 2002.
6) Rao, C. R., *Statistics and Truth: Putting Chance to Work, 2nd ed.*, World Scientific Publishing Co., New Jersey, 1997
7) <www.census.gov>
8) Peterson, I. "Census Sampling Confusion." *Science News Online*, v. 155 no.10 (March 6, 1999). <www.sciencenews.org/pages/sn_arc99/3_6_99/bob1.htm>
9) Ireland, R. "Board of Regents donate to Doyle." *University of Wisconsin-Oshkosh Advance-Titan*. April 28, 2004.

(Section 6: The St. Petersburg Lottery & The Problem of Points)

1) <http://www.imdb.com>
2) < https://www.boxofficemojo.com/release/rl3781789185/>

(Section 7: Shufflehall & How Many Tanks?)

1) Brodie, H. and R. Ruggles. "An Empirical Approach to Economic Intelligence in World War II." *Journal of the American Statistical Association*, Vol. 42, No. 237 (March 1947), 72-91.
2) Mosteller, F. *Fifty Challenging Problems in Probability with Solutions*. Addison Wesley, 1965. Republished by Dover, 1987.
3) Flaspohler, D. C. and A. L. Dinkheller. "German Tanks: A Problem in Estimation." *Mathematics Teacher*, Vol. 92, No. 8 (November 1999), 724-728.
4) AccuWeather.com <http://www.accuweather.com/us/wi/oshkosh/54901/forecast-month.asp?mnyr=7-01-2010>
5) Stahl, S. "The Evolution of the Normal Distribution." Preprint.
6) Hald, A., *A History of Probability and Statistics and Their Applications before 1750*, John Wiley and Sons, New York, 1990.
7) DeGroot, M. H. and M. J. Schervish, *Probability and Statistics, 3rd ed.*, Addison-Wesley, 2002.
8) College Board <www.collegeboard.com/sat/cbsenior/stats/stat001b.html>
9) ACT Research <www.act.org/research/briefs/2002-1.html>
10) Grierson, B. "The Hound of the Data Points." Popular Science, March 2003. <www.popsci.com>
11) Feynman, R. P., R. B. Leighton and M. Sands. *The Feynman Lectures on Physics, Volume I.* Addison Wesley, 1963.

(Section 8: The Illuminated Dartboard & Three Prisoners)

1) Rubel, L. H., "Good Things Always Come in Threes: Three Cards, Three Prisoners, and Three Doors," Mathematics Teacher, Vol. 99, No. 6, February 2006.
2) vos Savant, M., "Ask Marilyn," *Parade*, (a) September 9, 1990; (b) December 2, 1990; (c) February 17, 1991
3) Gillman, L., "The Car and Goats Fiasco," *Focus*, June-July 1991.
4) Greene, B., *The Fabric of the Cosmos: Space, Time, and the Texture of Reality*, Alfred A. Knopf, New York, 2004.
5) Arias, E., "United States Life Tables, 2002," *National Vital Statistics Reports*, Vol. 53, No. 6, November 10, 2004. Available from the U.S. Department of Health and Human Services, Center for Disease Control / National Center for Health Statistics. www.cdc.gov/nchs
6) Hald, A., *A History of Probability and Statistics and Their Applications before 1750*, John Wiley and Sons, New York, 1990.

7) Rao, C. R., *Statistics and Truth: Putting Chance to Work, 2nd ed.*, World Scientific Publishing Co., New Jersey, 1997.
8) Feller, W., *An Introduction to Probability Theory and Its Applications, Volume I, 3rd ed.*, John Wiley and Sons, 1968.
9) Gillman, L., "The Car and the Goats," *The American Mathematical Monthly*, Vol. 99, No. 1, January 1992, 3-7.

(Section 10: Seating Arrangements)

1) Miller, A. R., *The Cryptographic Mathematics of Enigma*, Center for Cryptologic History, National Security Agency, 2004.
2) Wilcox, J., *Solving the Enigma: History of the Cryptanalytic Bombe*, Center for Cryptologic History, National Security Agency, 2004.
3) Eastaway, R. and Wyndham, J., *Why Do Buses Come in Threes? The Hidden Mathematics of Everyday Life*, John Wiley and Sons, Inc., 1998.
4) Hawking, S., *A Brief History of Time: From the Big Bang to Black Holes*, Bantam Books, 1988.

(Section 11: MISSISSIPPI)

1) Brualdi, R. A., Introductory Combinatorics, 2nd ed., North-Holland, Elsevier Science Publishing Company, Inc., 1992.
2) Continental Airlines, Financial and Traffic Releases, Jan. 17, 2006 <www.continental.com>

(Section 14: Which Bag is Which?)

1) Weinberg, S. Dreams of a Final Theory: The Scientist's Search for the Ultimate Laws of Nature. Vintage Books, New York, 1993.
2) "colour blindness." Encyclopædia Britannica from Encyclopædia Britannica Premium Service. <http://www.britannica.com/eb/article?tocId=9024853>
3) webexhibits (an online museum). <http://webexhibits.org/causesofcolor/2.html>
4) "cystic fibrosis." Encyclopædia Britannica from Encyclopædia Britannica Premium Service. <http://www.britannica.com/eb/article?tocId=9028436>
5) Collins, Debra. "Genetics of Huntington disease". Division of Endocrinology, Metabolism and Genetics, University of Kansas Medical Center. <http://www.kumc.edu/hospital/huntingtons/genetics.html>

6) Pomfret, John. "In China's Countryside, 'It's a Boy!' Too Often". Washington Post Foreign Service. Tuesday, May 29, 2001; Page A01. <http://www.washingtonpost.com/ac2/wp-dyn/A77925-2001May25>
7) Banister, J. March 1999. "Son Preference in Asia -- Report of a Symposium." U. S. Census Bureau. <www.census.gov/ipc/www/ebspr96a.html>
8) U.S. Census Bureau, Population Division, International Programs Center. <http://www.census.gov/ipc/www/idbprint.html>
9) Aczel, A. D., Chance: A Guide to Gambling, Love, the Stock Market, and Just About Everything Else, Thunder's Mouth Press, New York, 2004.

(Section 15: ESP Testing)

1) Salsburg, David. The Lady Tasting Tea: How Statistics Revolutionized Science in the Twentieth Century. W. H. Freeman and Company, New York, 2001.
2) The Baseball Encyclopedia: The Complete and Definitive Record of Major League Baseball, 9th ed. Macmillan Publishing Company, New York, 1993.
3) Feynman, R. P. "Personal Observations on Reliability of Shuttle." Report of the Special Commission on the Space Shuttle Challenger Accident, Appendix F. National Aeronautics and Space Administration <http://history.nasa.gov/rogersrep/v2appf.htm>
4) National Aeronautics and Space Administration <http://www.nasa.gov/mission_pages/shuttle/main/index.html>
5) Stahl, S. "The Evolution of the Normal Distribution." Mathematics Magazine, 79:2 (2006), 96-113.
6) Campbell, N. A. Biology, 4th ed. Benjamin/Cummings, 1996.
7) Mendel, G. "Experiments in Plant Hybridization." 1866. Republished in: Newman, J. R. The World of Mathematics. Tempus Books of Microsoft Press. 1988.
8) Rao, C. R. Statistics and Truth: Putting Chance to Work, 2nd ed. World Scientific Publishing Co., New Jersey, 1997.

(Section 16: The Rhombus or the Trapezoid)

1) "Iraq drives Bush's rating to new low." CNN.com. March 14, 2006. <www.cnn.com/2006/POLITICS/03/13/bush.poll/>
2) Salsburg, David. *The Lady Tasting Tea: How Statistics Revolutionized Science in the Twentieth Century.* W. H. Freeman and Company, New York, 2001.
3) Baseball Almanac <http://baseball-almanac.com>

4) *The Baseball Encyclopedia: The Complete and Definitive Record of Major League Baseball*. Macmillan Publishing Company, New York.
5) Nielsen Media Research <www.nielsenmedia.com>
6) Barnes, B. "For Nielsen, Fixing Old Ratings System Causes New Static." *The Wall Street Journal*. September 16, 2004.

(Section 17: Fingerprints)

1) "fingerprint." Encyclopædia Britannica. 2005. Encyclopædia Britannica Premium Service17 June 2005 <http://www.britannica.com/eb/article?tocId=9034291>
2) SCAFO Online Articles, originally appearing in the August 1987 issue of *Identification News* <http://www.scafo.org/library/110203.html>
3) Gigerenzer, G. *Calculated Risks: How to Know When Numbers Deceive You*. Simon & Schuster, New York, 2002.
4) Begley, S. "Despite Its Reputation, Fingerprint Evidence Isn't Really Infallible." *The Wall Street Journal*. June 4, 2004.
5) "Testing Yourself for HIV-1, the Virus that Causes AIDS - Home Test System Is Available."U.S. Food and Drug Administration, Center for Biologics Evaluation and Research. <http://www.fda.gov/cber/infosheets/hiv-home.htm>
6) "A Glance at the HIV/AIDS Epidemic." Department of Health and Human Services, Centers for Disease Control and Prevention. <http://www.cdc.gov/hiv/pubs/Facts/At-A-Glance.htm#1>
7) *People v. Collins* 68 Cal. 2d 319, 438 P.2d 33, 66 Cal.Rptr. 497 (1968)
8) "What You Need to Know About Drug Testing in Schools." Office of National Drug Control Policy. <www.whitehousedrugpolicy.gov/pdf/drug_testing.pdf>
9) "Fact Sheet: Drug Testing in the Criminal Justice System." March 1992. U. S. Department of Justice, Office of Justice Programs, Bureau of Justice Statistics.
10) Visher, C. November 1991. "A Comparison of Urinalysis Technologies for Drug Testing in Criminal Justice." National Institute of Justice Research Report. U.S. Department of Justice, National Institute of Justice.
11) National Survey on Drug Use & Health. 2003. U.S. Department of Health and Human Services, Office of Applied Studies.
12) U. S. Census Bureau. <www.census.gov/popest/estimates.php>
13) Johannes, L. "Study Finds MRIs Improve Breast-Cancer Detection." *The Wall Street Journal*. July 29, 2004.
14) Liberman, L. "Breast Cancer Screening with MRI -- What Are the Data for Patients at High Risk?" *The New England Journal of Medicine*. July 29, 2004. Vol. 351, Iss. 5, p.497-500.

Glossary

Availability Fallacy: The tendency to make decisions regarding chance based on the how easily certain events can be called to mind.

Bar graph: a data display in which the vertical axis represents a count (frequency) or a percent (relative frequency) and the horizontal axis represents values of a categorical variable (like colors), values of a numeric discrete variable (like shoe size), or values of a continuous variable (like heights).

Base Rate Fallacy: The tendency to make decisions regarding chance that neglect the rates at which relevant characteristics appear in the population.

Bayes' Theorem: A theorem explaining how to invert conditional probabilities.

Binomial distribution: Suppose there is a set of repeated independent trials, each trial having two possible outcomes (we'll call them *success* and *failure*), and the probability of success is the same on each trial. The *binomial distribution* is the probability distribution of the *total* number of successes.

Blind: A clinical study is blind if the participants do not know whether or not they have received the treatment.

Categorical Data: Data for which it makes sense to place an individual observation into one of several groups (or categories).

Chance Error: Error in sampling due simply to the fact that a sample is not exactly the same as a population due to chance alone.

Census: Any data collection method in which data is collected from each and every member of the population.

Central Limit Theorem: A theorem which says that for repeated, independent trials, each with the same underlying probability distribution, the sample mean has a *normal* (bell-shaped) probability distribution.

Clinical study: Any study designed to determine a cause and effect relationship between two variables.

Combination (of a set): A subset of the set. Often a *combination* is contrasted with a *permutation*, in that a permutation of a set refers to an ordered arrangement of elements, but a combination just refers to the collection of elements, without regard to order. The number of combinations of size k, chosen from a set of size n, is the same as the number of permutations of a multiset containing two distinct elements, one of which has *k* copies (appears *k* times) and the other of which has the remaining *n-k* copies.

Confidence interval: An interval around a statistic, chosen to exclude those possible values of the parameter for which the observed statistic would have been unlikely to occur.

Confounding variable: Any variable that was not controlled in the experiment and may have caused the observed effect.

Conjunction Fallacy: The tendency to make decisions regarding chance without regard to the fact that A and B both occurring is less likely (or equally likely) than A occurring alone.

Control group: In a controlled study, the control group is a group of participants who does not receive the treatment under study and is used for comparison with the treatment group.

Correlation coefficient (r) is a statistic that measures the strength of the linear relationship between two numerical variables.

Data: Any information collected from a sample.

Distribution of data: the values the data takes along with the frequency (or relative frequency) of those values.

Disjoint (events): Two events are disjoint if they have no outcomes in common.

Double-blind: A clinical study is double-blind if neither the participants nor the researchers know who has received the treatment (until after the study is over).

Empty set: The set with no elements, often denoted by { }.

Equally likely outcomes: All the outcomes of a random experiment are said to be equally likely if, in the long run, they all occur with the same frequency (they all have the same probability of occurring).

Event: a subset of the sample space of a random experiment.

Expected value (of an observation whose possible outcomes are numerical values): the average of the values that could occur, weighted according to the underlying probability distribution.

Explanatory variable is one that is used to explain variations in another variable, called the response variable.

Extrapolation: Making a prediction outside the bounds of the data.

False negative: A test result which fails to identify the condition that is being tested for, although the condition is present.

False positive: A test result which falsely indicates the presence of the condition that is being tested for, although the condition is not actually present.

First quartile (Q_1): the median of the lower half of the data (the data positioned strictly before the median when the data is placed in numerical order).

Five Number Summary: of numerical data consists of the following five statistics: Minimum observation, Q_1, Median, Q_3, Maximum observation.

Gambler's Fallacy: The tendency to make decisions regarding chance based on the (mistaken) belief that after a long string of losses, a win becomes more likely. (This is a type of representativeness fallacy.)

Histogram: A data display in which the vertical axis represents a *rate* and the horizontal axis is actually the *x*-axis -- in other words, it is a continuous piece of the real number line and as such it contains all real numbers in an interval. Possible values of the variable are broken into interval classes along the horizontal axis; a bar is drawn above each class, with the *area* of the bar representing the frequency or relative frequency of that class. (*Note*: some texts use the terms *histogram* and *bar graph* interchangeably. We will not do so in this text.)

Hypothesis test: A test which begins with a "null" hypothesis regarding a parameter, and then evaluates the likelihood of the observed statistic, assuming the hypothesis to be correct. This helps the observer decide whether to accept or reject the null hypothesis.

Independent events: Two events are said to be independent if knowing that one occurs (or knowing it does not occur) does not change the probability of the other occurring.

Interpolation: Making a prediction within the bounds of the data.

Inter-Quartile Range (IQR) = $Q_3 - Q_1$. It is the spread of the middle half of the data.

Law of large numbers (for means): A theorem which says that in repeated, independent trials, each with the same underlying probability distribution, the chance that the sample mean (the average of the outcomes) differs by any given amount from the population mean approaches zero as the number of trials increases.

Law of large numbers (for proportions): A theorem which says that in repeated, independent trials, each with the same probability of success, the chance that the proportion of successes differs from the probability of success approaches zero as the number of trials increases.

Least Squares Regression Line: the line that minimizes the sum of the squares of the vertical distances of the points to the line on a scatterplot.

Linear Correlation: two numerical variables are said to have a linear correlation if they have a strong *linear* association.

Mean (of a population): For a population consisting of numeric values, the mean is the weighted average value; essentially, this is the sum of all possible values, each multiplied by its probability of occurrence.

Mean (of a sample): For a sample of numeric data, the mean is the average value, that is, the sum of all the observed values divided by the number of observations.

Median: The median of a set of numeric data is the middle value when the data is placed in numerical order. In the case where there are two "middle" values, the median is any number between those two (often taken to be the average of the two).

Mode: The mode of a set of data is the data value that occurs most frequently.

Multiplication principle: If you have A different ways of doing one thing and B different ways of doing another (and one choice does not affect the other) then the total number of ways to do both A and B is $A \times B$.

Multiset: A collection of elements, some of which may be copies (repeats) of other elements in the collection. Two multisets are distinguished from each other not only by which elements they contain, but also by how many times each element appears.

Mutually Exclusive Events: events which share no common outcomes. (They are disjoint sets.)

Negative association: Two variables measured on the same individuals have a negative association if increases in one variable tend to correspond to decreases in the other variable.

Nonresponse bias: Biased information that results from the fact that some groups in the sample were more likely to respond than others.

Numerical Data: number data for which it makes sense to perform arithmetic operations (such as averaging).

Outcome: Any one thing that could happen in a random experiment.

Outlier: a specific observation that lies well outside the overall pattern of the data.

Parameter: A representative number computed from an entire population (e.g. a population *mean* is a parameter).

Permutation (of a set): An ordered arrangement of the elements of the set.

Permutation (of a multiset): An ordered arrangement of the elements of the multiset, where the repeated elements are regarded as indistinguishable from each other.

Placebo: a fake treatment given to the control group in a clinical study.

Population: the largest group about which the researcher or surveyor would like information.

Positive Association: Two variables measured on the same individuals have a positive association if increases in one variable tend to correspond to increases in the other variable.

Probability of an event: the likelihood of that event occurring. Often this can be thought of as the percentage of times the event would occur if a random experiment were performed a very large number of times.

Range = maximum observation – minimum observation.

Random Experiment: any experiment where the outcome depends on chance and cannot be known beforehand. While, in a random experiment, the specific outcome cannot be known, there is nonetheless a regular distribution of outcomes after a very large number of trials.

Randomized-controlled: a clinical study is said to be randomized-controlled if participants are assigned to treatment and control groups at random.

Ratio: a comparison of two counts or measures that have the same unit.

Representativeness Fallacy: The tendency to make decisions regarding chance based on the (mistaken) belief that even small samples should be representative of population.

Response Rate: The ratio of the number of respondents (people who actually took part in the study) to the number who were invited to participate.

Response variable: a variable whose variation is explained by another variable (called an explanatory variable).

Sample: a subset of a population from which data is collected

Sample bias: Biased (inaccurate) information that results from a poorly chosen sample.

Sampling error: the difference between the value of a parameter (such as the population mean) and the value of the statistic that represents it (such as the sample mean). Sampling error can include chance error, sampling bias, and nonresponse bias.

Sampling rate: [(number of elements in the sample) ÷ (number of elements in the population)]

Sample space: The sample space of a random experiment is the set of all possible outcomes.

Scale factor, a value by which we multiply each original dimension to find the new lengths.

Scatterplot: A display that shows the relationship between two numerical variables measured on the same sample of individuals. The values of one variable are shown on the horizontal axis, and values of the other variable are shown on the vertical axis. Each individual is plotted as a point representing an ordered pair (variable 1, variable 2).

Self-selected sample: a sample in which respondents volunteered to participate.

Set: A collection of distinct elements. Two sets are distinguished from each other only by which elements they contain.

Similar: Two geometric figures are similar if there is some scale factor we could use to make them coincide. (Similar figure have the same shape, but are not necessarily the same size.)

Simple Random Sample (SRS): A sample in which every subset of the population has the same chance of being in the sample as any other subset of the same size.

Skewed left: A distribution of data is skewed left if it has an asymmetric tail to the left.

Skewed right: A distribution of data is skewed right if it has an asymmetric tail to the right.

Slope (m): the slope of a line is the change in y for a one unit change in x

Standard deviation (of a sample): The standard deviation of a sample is a statistic that measures spread around the mean. Where n is the size of the sample (the number of observations), $\bar{x}$ is the sample mean, and each x_i is an individual observation, the standard deviation of the sample is

$$\sqrt{\frac{(x_1 - \bar{x})^2 + (x_2 - \bar{x})^2 + \cdots + (x_n - \bar{x})^2}{n-1}}.$$

Stratified random sampling: Random sampling in stages (or layers or *strata*).

Statistic: Any numerical information computed from a sample (e.g. a sample *mean* is a statistic).

Stem and leaf plot (or simply **stemplot**): is a listing of all the data typically arranged so the tens place makes the stem and the ones are the 'leaves.'

Survey: Any data collection method in which data is collected from just a subset of the population.

Symmetric Distribution: A distribution that is symmetric about the median.

Third quartile (Q_3): the median of the upper half of the data (the data positioned strictly after the median when the data is placed in numerical order).

Treatment group: in a controlled study, the treatment group refers to the participants who are given the treatment that is being tested to see if it causes a certain effect.

Uniform distribution: a probability distribution in which every outcome has the same probability of occurrence.